AGGRESSIVE BEHAVIOR OF PEOPLE WITH AUTISM AND OTHER INTELLECTUAL AND DEVELOPMENTAL DISABILITIES

AGGRESSIVE BEHAVIOR OF PEOPLE WITH AUTISM AND OTHER INTELLECTUAL AND DEVELOPMENTAL DISABILITIES

Edited by

LAUREN CHARLOT
UMass Chan Medical School, Worcester, MA, United States

ANGELA HASSIOTIS
University College London, London, United Kingdom

ELSEVIER

ACADEMIC PRESS
An imprint of Elsevier

Academic Press is an imprint of Elsevier
125 London Wall, London EC2Y 5AS, United Kingdom
50 Hampshire Street, 5th Floor, Cambridge, MA 02139, United States

ISBN: 978-0-443-26552-5

For information on all Academic Press publications visit our website at https://www.elsevier.com/books-and-journals

Publisher: Megan Ball
Acquisitions Editor: Megan McManus
Editorial Project Manager: Swadha Puri
Production Project Manager: Gomathi Sugumar
Cover Designer: Mark Rogers

Typeset by Integra Software Services

Contents

About the authors

Lauren Charlot (co-editor), PhD, is a developmental psychologist and an assistant professor in the Department of Psychiatry at UMass Chan Medical School, where she specializes in autism spectrum disorders (ASD), intellectual and developmental disabilities (IDD), and cooccurring psychiatric and behavioral health challenges. With over 30 years of clinical, research, and program development experience, Dr. Charlot has conducted seminal research on psychiatric presentations, mood and anxiety assessment tools (including work validating the MASS interview), and patterns of psychiatric hospitalization among individuals with IDD and ASD. She has designed, directed, and evaluated innovative inpatient and outpatient neuropsychiatric services for people with complex needs, emphasizing *person-centered*, multidisciplinary, trauma-informed approaches to care and reducing reliance on restrictive interventions. Dr. Charlot has lectured widely in the United States, Canada, and Europe, and contributed to clinical practice guidelines addressing comprehensive mental health assessment and treatment in IDD/ASD populations.

Angela Hassiotis (co-editor) is a Professor of the Psychiatry of Intellectual disability at the UCL Division of Psychiatry and an honorary consultant psychiatrist at the award-winning Camden Intellectual Disability Service. Her main research interest is the evaluation of interventions for mental ill health including behaviors of concern in people with intellectual disabilities across the lifespan. She has run several clinical trials and epidemiological studies and has authored over 300 publications. She supervises MSc and PhD students and lectures nationally and internationally. She has received the NADD Stephen Reiss Research Award (2020) and the Frank Menolascino Award for Excellence (2023). She was the editor-in-chief of the *Journal of Mental Health Research in Intellectual Disabilities*, is one of the medical advisors for research at the Baily Thomas Charitable Foundation, UK, and a member of the Academic Psychiatry Faculty, RC Psych, UK. She has played a prominent role in shaping UK policy and clinical guidance related to intellectual disability and mental health, collaborating closely with the NHS and national bodies. Through her mentorship, publications, and advocacy, Dr. Hassiotis has been a key figure in advancing specialized, humane, and effective mental health services for people with IDD across the lifespan.

Karyn Harvey, PhD, is a psychologist clinician, author, and trainer with more than three decades of experience in the field of IDD. She is widely recognized as an expert on trauma of people with ASD/IDD. She holds a PhD in Applied Developmental Psychology from the University of Maryland and has devoted her career to advancing trauma-informed care for people with IDD, emphasizing how unrecognized trauma contributes to behavior and mental health challenges. Dr. Harvey is the author of several influential books, including *Positive Identity Development*, *Trauma-Informed Behavioral Interventions*, and *Trauma and Healing in the Lives of People with Intellectual Disability*, which serve as key resources for clinicians, caregivers, and direct support professionals. She maintains a private practice offering individual and group therapy for individuals with IDD and regularly conducts trainings across the United States and Canada on trauma-informed support, leadership, and therapeutic interventions. Her work champions safety, empowerment, and connection as foundational elements for healing and meaningful quality of life for people with autism and IDD.

J. Leonard Peebles, PhD, is a retired forensic psychologist whose main area of interest is aggressive behavior in various target populations. He has extensive experience in the assessment of aggression and served as an expert witness in over 300 hearings in District and Superior Courts in Massachusetts. Dr. Peebles has done numerous presentations and has written on the topic of aggression. After leaving forensics work, he developed a special interest in trauma experienced by crisis responders and individuals who have been repeatedly faced dangerous crisis situations involving aggression or the threat of aggression. He has often worked closely with colleagues and other professionals who have faced trauma in the course of their work, providing supervision and support. Finally, he worked in child protective services performing numerous initial assessments in Alaska. In addition to forensics, he has presented and written about education and training for caregivers of people with IDD with aggressive behavior. Dr. Peebles and a colleague developed a highly respected, specialized 20-h course for direct support providers, emphasizing ways to prevent and more calmly and efficiently support people with IDD who engage in behaviors that are dangerous to people around them and themselves. His study on this topic showed that people who took the 20-h course demonstrated significantly more confidence in their abilities to respond to incidents of aggressive, dangerous behavior.

I. Leslie Rubin, MD, is an associate professor in the Department of Pediatrics at Morehouse School of Medicine, adjunct associate professor in the Department of Pediatrics at the Emory University School of Medicine, and medical director of the Rubin Center for Autism and Developmental Pediatrics. In addition, he is the director of the Break the Cycle Program for the Southeast Pediatric Environmental Health Specialty Unit at Emory University School of Medicine and is the founder of Break the Cycle of Health Disparities, Inc., a nonprofit organization dedicated to exploring the social, economic, and environmental determinants of health and working with university student to develop projects that will break the cycle of health disparities. As a developmental pediatrician, he is clinically active in caring for children with developmental disabilities and their families in his private practice, as well as directing interdisciplinary clinics for children with autism and cerebral palsy at CHOA Hughes Spalding, and consulting to Hall County Children's Medical Services and the Emory Neurodevelopmental Exposure Clinic. He is the primary editor of *Health Care for People with Intellectual and Developmental Disabilities across the Lifespan* (Springer, 2016). He also started the annual Break the Cycle of Children's Environmental Health Disparities program in partnership with the Southeast PEHSU to raise awareness among university students of children's environmental health disparities and to cultivate future leaders to address these challenges. The program received the Children's Environmental Health Excellence Award from the US EPA in 2016. Dr. Rubin served for 6 years on the Board of Scientific Counselors to the EPA ORD and serves on the National Advisory Council for Children and Disasters. His personal mission is to work in collaboration with others on reducing children's health disparities and promoting health equity for all children, locally, nationally, and globally in the context of environmental and social justice and in the face of climate change.

Foreword

Aggressive behaviors among people with autism spectrum disorder and other neurocognitive/intellectual and developmental disabilities remain one of the most misunderstood and difficult areas in clinical practice at all levels of health care (ambulatory, access to specialists, procedures, emergent care, and inpatient medical and behavioral health services. Aggression, when part of the presentation, is often the most challenging and urgent one faced by clinicians, caregivers, families, human service systems, and habilitative supports. A history of/or current aggression often impacts assessment, evaluation, treatment, and access to services in health care systems and the community.

What makes this book especially valuable is its clear recognition that aggression is a symptom and communicates something—it is not a diagnosis, and nonspecific treatment is problematic at all levels. Aggression is often complex, contextual, and cannot usually be understood through any single lens. This work and the chapters that follow encourage clinicians and readers to move beyond simple reactive behavior management and to ask deeper questions: What is the person communicating with their behaviors? What are all the sources of distress or frustration that must be identified, understood, and managed simultaneously that are driving the behaviors? How do developmental factors, trauma, medical conditions, co-occurring mental health challenges, and environmental factors, such as life's experiences and context the behavior occurs in, shape the behaviors that are seen or reported?

A particular strength of this work is its insistence on seeing the whole person and understanding their experience of the world—interventions often become clear once this is accomplished. Throughout the work, aggressive behaviors are approached not merely as a target for reduction but also as a likely expression and communication of distress, frustration, fear, illness, unmet needs, and disrupted well-being.

Aggression is often inappropriately attributed to the person's neurodevelopmental disability—other etiologies that may be contributing to the individual's distress and often are not considered or addressed (see DA Pinal et. al. 2022). It often places enormous strain on the individual, their families, their caregivers, and systems of care and services. It often presents with and is associated with crisis responses, exclusion, stigma, and reduced quality of life, and amplifies the disparity of access and care issues already present.

In this volume (book), Lauren Charlot and Angela Hassiotis bring together many important bodies of knowledge and practice wisdom focused on the many possible influences and factors that may contribute to aggression in people with autism and other intellectual and developmental disabilities and provide a blueprint for understanding and approach. The chapters in this book reflect a central and compelling premise: aggression is rarely a simple phenomenon and is seldom explained by a single cause. Rather, it emerges through the interaction of developmental differences, distress caused by missed physical health conditions, co-occurring mental health challenges, the impact of cumulative trauma histories, often challenges with communication, contextual environmental stressors, caregiving environments, service responses, and a lack of service access.

This perspective, if embraced, can have profound implications for proactive responses. It effectively shifts the focus from reactive management to thoughtful formulation based on understanding, from narrow interventions to integrated multidisciplinary care, and from management and control alone to the promotion of safety, dignity, wellness, and an improved quality of life. If these principles/knowledge are incorporated, it will aid people in being well, feeling understood, feeling more in control, having agency, and receiving the health care and supports that truly fit their needs.

Scott. R. Stiefel
Medical Director of Behavioral Health and Developmental
Disabilities Services, NYC Health and Hospitals,
New York, NY, United States.

Preface

Michael was about 19 years old and had just been learning about the label that was used to describe him clinically at the time, "Asperger's syndrome." The interview was done some years ago. He agreed to let us use some of his comments and insights.

Asked what he thought about being diagnosed as having Asperger's syndrome. He said, "Some are also misdiagnosed … I was misdiagnosed for that kind of stuff in sixth grade … There was a bit of a misunderstanding. I got arrested. It was difficult but hey, I got through it … I wasn't listening to the Principal. I was in the Principal's office, and I wouldn't stay seated and I wouldn't stop making noise."

The interviewer noted how he did *so* well after he changed schools. She shared how he had been diagnosed with Conduct Disorder and was sent home many times due to being loud and disruptive, and fears of his becoming aggressive. His mother advocated for him and got him moved to a different school district (a difficult accomplishment). The new school recognized him as having a form of autism, and accommodations were made. Michael added, "In High School, I didn't even expect it, I got High Honors."

The interview went on, and Michael stated, "I would do whatever they wanted me to do if I just knew what it was." He admitted he did not always understand what was socially expected. He described kindness to people who teased him. He described how his "meltdowns" would happen when he was anxious because others got too close to him (into his personal space). It was agreed by the end of the interview that he is a very sweet, caring person. He was helped by attending social skills building groups and was no longer at risk of engaging in aggressive outbursts. The autism label doesn't mean much anymore. He is just a nice man who works, drives his car, and has an apartment and a cat. He is amazing at Sudoku and trivia too (you want him on your team).

Acknowledgments

From L. R. Charlot: I am especially grateful to **Jean Frazier, MD,** and **Joni Beasley, PhD,** both of whose mentorship and leadership have had a lasting influence on my professional development.

We wish to express our gratitude to the young man whose art was used for the cover of this book. Mr. Alfred Brown III is a very talented person with autism. He has been supported in his amazing creative endeavours not only by his family, but by a very special agency serving people with autism and other intellectual and developmental conditions in Erie, PA, USA. **Barber National Institute** demonstrates how supporting people to build on their strengths in a positive, caring environment helps to produce happy people who have a sense of purpose. We hope this perspective is reflected in our book. Essentially, we believe that evidence informed focus on wellness can prevent aggressive behaviour and improve the quality of life of people we serve. When I met the artist, I was so impressed with his big mile, his clear pride in his art, and his friendly nature as well as the equally warm approach of the staff.

CHAPTER 1

A final common pathway: Multiple influences on aggression

Lauren Charlot[a] and Angela Hassiotis[b]
[a]UMass Chan Medical School, Worcester, MA, United States
[b]University College London, London, United Kingdom

The goal of the book is to describe strategies for the assessment and treatment of aggressive behavior exhibited by people with autism spectrum disorder and other intellectual and developmental disabilities (ASD/IDD). Most of the assessment and treatment suggestions made here pertain to people with neurodevelopmental conditions. In some instances, the suggested approaches can be helpful whether a person has met a formal definition of intellectual disability or not, but they are more directed at the clinical care of people who display developmental disabilities generally associated with a neurodevelopmental condition or etiology that makes it necessary to have caregivers (paid or family) or some form of disability-driven supports. We point this out given a basic dilemma writers have noted in deciding how to identify the people about whom they are concerned. (see excellent discussion by Marc Tasse: Tasse et al., 2024). We refer to people who have an ASD as people with autism, or use a person-first approach, recognizing there are different ways people prefer to make such references.

We have not dedicated chapters to specific aspects of practice such as confidentiality, consent to treatment, capacity assessment, and use of restrictions. These are all essential parts of good and ethical care, and we assume that they will be covered by statewide or federal legislation as it applies to each case. All practitioners must ensure that identifiable information about a person is not shared unless there is explicit agreement to do so by the person themselves, and that where treatment is to be started or existing treatment is to be changed, informed consent is sought from an adult with ASD/IDD, those with parental authority if the person is a child, or in the best interests of the person if they are an adult but lack the capacity to consent. The latter involves evidence that the person understands the purpose of the treatment, can retain information about the treatment, what may be the consequences of not receiving the recommended treatment,

Aggressive Behavior of People with Autism and other Intellectual and Developmental Disabilities
https://doi.org/10.1016/B978-0-443-26552-5.00001-1

and can make a final decision about it. It is very important to bear in mind that capacity is decision-specific and that the ability to reach a decision can fluctuate significantly, especially in the presence of a mental disorder.

We believe that aggressive behavior is an important clinical practice topic and is a common presentation among people with ASD/IDD. Multiple studies show that it adds to the disease burden across the lifespan, increases the odds of a mental health crisis and hospitalization, leads to the use of restrictive practices, stigma, and exclusion, and is costly for services. No single intervention adequately manages aggressive behavior, and good practice in supporting individuals who display the behavior is variable within and across countries globally.

In the course of addressing the occurrence of aggressive behavior, the suggestions are also relevant to other significant behaviors that challenge within the domain of aggressive behaviors, including self-injurious acts and aggressive–destructive behavior. Aggression and other significant behavioral concerns are often the basis for referrals to *behavioral health* or *psychiatric services*. Though varied models and approaches each have significant value in the work to understand and treat aggressive behavior in the population addressed here, it seems clear that more can be done to improve care. A brief overview of the most common variables found to influence the occurrence of aggressive behavior in this population is provided. The importance of developing a case formulation that integrates a developmental perspective with the usual biopsychosocial multidisciplinary assessment is emphasized. Also, the need for a contextual framework in conducting positive behavior support and treatment planning using multiple modalities is urged as a way to more effectively promote wellness and reduce aggressive behavior (Kalb et al., 2012).

In addition to citing some of the related research, we have contributed ideas, tools, and suggestions based on our personal professional experience in multiple settings within which the care and treatment were developed to help individuals with ASD/IDD and co-occurring aggressive behaviors. Dr. Charlot is a developmental psychologist who has designed and run crisis services, short-stay crisis houses, outpatient psychiatric clinics, a locked psychiatric inpatient unit specializing in the care of people with ASD/IDD, and a specialist multidisciplinary IDD/MH evaluation team. She is using lessons learned while working to help people served in these settings. Professor Hassiotis is a consultant psychiatrist practicing in the UK National Health Service and has extensive experience in the management of aggressive behaviors in the population with ASD/IDD and has developed and tested interventions for different psychiatric disorders in this population.

In many cases, aggression may best be understood as a *final common pathway* for distress among people who have a limited repertoire for expressing feelings (Charlot et al., 1993). The goal of assessment is to understand what factors contribute to the journey along these pathways. Many experts have described the need for a *bio-psycho-social* understanding of the roots of mental illness. This is also true for understanding aggressive behaviors displayed by people with ASD and other forms of IDD. In the work presented here, we add a *developmental domain*. Developmental approaches can be integral to understanding aggressive acts of people with a neurodevelopmental disorder. Applied behavioral analysis is also a frequently used area of assessment and treatment for aggressive behaviors. Suggestions for using behavioral assessment techniques with more focus on contextual influences on aggressive behavior are presented. The proposed contextual assessment and treatment (CAT) approach aims to provide a means for examining contextual influences on mood and behavior and encompasses trauma-informed care. Many factors act as "setting events" and add risk for aggressive acts. Accurate identification of contextual influences can inform more effective, preventive, and multidisciplinary assessment and treatment planning.

Comprehensive case formulation is an antidote to checklist approaches in typical functional assessments that focus only on broad examples of consequent events maintaining challenging behaviors. A case formulation may also be important in achieving accurate psychiatric diagnoses. An integrated case formulation addressing these varied "influences" informs a more comprehensive approach to care and treatment. The varied areas of influence of most importance will be reviewed in more detail in subsequent chapters and are briefly introduced below. Developing a comprehensive psychiatric case formulation for a person with ASD/IDD who presents with aggressive behavior can be enhanced by considering that unique individual's neurodevelopmental profile.

A critical step in the assessment of behaviors that challenge and mental health disorders is to know what being *well* looks like in the person one is evaluating. When a person shows an increase in aggressive behavior or the onset of such behavior, it is important to understand the extent to which these behaviors represent a departure from usual behavior for this unique individual. Symptoms of psychiatric disorders are almost always defined as representing a departure from "usual behavior." This generally means usual for people of a given age, in a certain culture, or with a particular background. For our purposes here, people with a similar neurodevelopmental profile, rather than a certain age, should be considered as a reference group for what may be expected. Ideally, we want to understand the individual person's neurocognitive and social-emotional developmental profile. Each person with ASD or other IDD will have a unique profile for what is "usual behavior," from which we can define what is a departure (what is unusual or atypical). We should also consider how people from a similar cultural background and with similar financial, residential, and service histories might be expected to present in the circumstances that bring them to clinical attention. How one defines psychopathology is critical to how interventions are developed. In many situations, goals are set to lower the frequency or intensity of behaviors that are challenging. However, ideally, the goal should be for the person to be well, with an expectation that when people are well, they are less likely to engage in aggressive acts.

Knowing that a particular behavior that challenges is maintained by some form of "attention" is useful but minimally so without a broader look at contextual factors. Knowing details of the contexts in which the behavior that challenges is more likely is most helpful in developing a preventive plan of approach. For example, if a person is seeking *attention* today because of an impacted wisdom tooth, knowing this is the primary factor influencing agitation and aggressive acts today is critical to an effective intervention. In such cases, some psychotropic medications and certain operant conditioning tactics may reduce aggressive acts *transiently*. Any reduction in the negative targeted behavior is often seen as confirming the intervention was effective, despite the problem reemerging. In a similar way, labeling moodiness as evidence of psychiatric illness may be seen as confirmed by the effects of medication modifying emotional expression. Finding and securing treatment for the impacted wisdom tooth, however, is a better approach for obvious reasons. If, as suggested, the aggressive acts are nonspecific evidence of a problem (not *what* problem), each occurrence may need a new look. Further, even if we agree that somebody was seeking "attention," it is important to know why and why now. For example, what kind of "attention" was the person seeking? All attention is not equal. Many cases of individuals exhibiting aggressive behavior seen as aimed at or maintained by attention have occurred while the person was supervised by one, two, or even three staff persons. In some instances, the aggression is being driven by an acute psychiatric issue, so this needs further evaluation. In many cases, "escape" is a maintaining factor driven by a history of trauma. Trauma triggers may provoke either fight (aggression) or flight (trying to run away from supervision). The developmental contextual assessment and treatment approach provides ways to identify influencing factors. How the use of this approach, with traditional PBS strategies included, can inform interventions within a multidisciplinary treatment plan is presented.

The key in all assessments, including psychiatric ones, is knowing that the behaviors themselves may only be a surface manifestation of distress or frustration. They tell us only that something is wrong, but not what is awry. The specific forms aggressive acts take are most often shaped by the person's neurodevelopmental profile and personal history. These are people who, by virtue of their disability, have a *limited behavioral repertoire*. More simply put, people with ASD/IDD often have fewer ways to demonstrate distress than others with a neurotypical developmental history. Much of the time in a psychiatric or psychological visit will be spent elaborating details of what the aggressive behaviors look like—their form, frequency, and intensity—as well as their impact on others. Here, we suggest ways to take into account neurodevelopmental and contextual influences on psychiatric disorders. This includes, for example, considering when aggressive acts are simple manifestations of frustration for a person with ASD who lacks understanding of social rules and expectations, and when these acts are primarily a manifestation of an acute psychiatric episode (or both).

In Chapter 2, we examine the variable nature of neurodevelopmental profiles of people with ASD and IDD. Understanding each person's unique manner of perceiving their world, processing information, and problem-solving helps us identify factors that impact the occurrence of aggressive behavior. To identify root causes of aggression, it is important to know how a person functions when well and at their best. Often, people react to stress in ways tied to their developmental capacities and challenges. For example, some young children with a neurotypical developmental history become quite agitated when anxious, while this is less common for older individuals. Adults with IDD may respond to stress more similarly to how a younger neurotypical person would, in part due to similarities in how information is processed. Additionally, people will respond best to interventions aimed at reducing aggression that are framed by their individual profile of strengths and weaknesses. This includes the person's general means of information processing and planning, but also expressive-receptive verbal abilities, visuo-spatial processing, and social-emotional development. The ways in which assessment of aggressive behavior can be enhanced by including information found in IQ testing, adaptive behavior assessments, autism-specific assessments, and evaluations of components of executive function are discussed. Approaches developed to reduce aggression can be formulated to reflect findings about how individuals learn best. For example, if a person shows relative strength in visual processing and weakness in verbal arenas, we will include the use of visual materials in our treatment plan. Or, a person who has ASD and an average or above-average FSIQ but is very challenged in understanding social cues may need a plan crafted to enhance these weaker capacities. When people with IDD are unable to convey what they need and hope for, or have difficulty connecting with others, aggressive acts may be more probable. The use of a developmental framework enhances behavioral and psychiatric assessment and provides an important addition to a bio-psycho-social multidisciplinary model for understanding aggressive behavior (Matson and Adams, 2014).

In Chapter 3, we look closely at common and uncommon medical conditions and aggressive behavior in this population. Numerous medical problems have been found to impact the mood and behavior of people with ASD and other IDD. There are considerable data showing that people with these disabilities have more health problems and continue to face access issues. In this chapter, the most frequent medical problems found to contribute to dysphoria, and as a result, to aggressive behavior, are described. Tips to ensure frequently missed medical problems are captured are also provided. Suggestions for surveillance and treatment are offered when these differ for individuals with IDD. For many medical problems, treatment is the same for all people, and the primary concern is not to miss them when they are affecting a person with ASD/IDD (Rubin et al., 2016).

In Chapter 4, *Genotypes, Phenotypes, and Aggressive Behavior*, some of the most common genetic anomalies associated with ASD or other IDD are discussed, with a focus on their contribution to the risk of aggressive behavior. Behavioral phenotypes, environmental transactions, and epigenetics are also examined. Down syndrome, Fragile X, Prader-Willi, Cornelia de Lange, and 22q11.2 deletion syndromes are each briefly described, along with the risk for aggression conferred by their various phenotypical manifestations (Rubin et al., 2016).

> *The adult being is an emergent entity who must be understood at his own level and in his own totality. The truly salient issues are malleability and flexibility, not fallacious parsing by percentages. A trait may be 90 percent heritable, yet entirely malleable.*
>
> **Stephen J. Gould**

In Chapter 5, we examined what constitutes a psychiatric syndrome and highlighted the need for a comprehensive model to support accurate psychiatric diagnoses in the population. Accurate identification of psychiatric disorders is essential for effective treatment, including addressing aggressive behaviors that are sometimes a downstream effect of an acute mental health condition. Assessment of individuals with ASD or other IDDs can be challenging. Tips to improve diagnostic accuracy by studying the clinical course and history of each patient's presenting challenges, in addition to listing symptoms, are shared. Challenges and ways to overcome barriers are suggested. Given that symptom manifestations vary in neurotypically developing youth of different ages, how developmental features might do the same in people with ASD and other IDDs is also explored (Charlot et al., 2022).

One-page summaries for three major symptom clusters and syndromes are provided. The topic of trauma is addressed in a separate chapter due to the extremely high rates in the population and because of the hypothesis put forth here that trauma is a driving force in much of the aggressive behavior we see. The goal in sharing one-page summaries is to help clinicians avoid problems with overdiagnosing some syndromes while missing others. It is hypothesized that inaccurate psychiatric labels can cause ineffective treatments, often inadvertently leading to ongoing aggression. The one-page summaries provide important clues to when certain suspected diagnoses are unlikely and signs that indicate the disorder is probably occurring. The tips provided are based on recognition of developmental effects and challenges when informant reports are a main source of clinical data:

- Psychotic disorders
- Bipolar disorder
- Anxiety disorders

Trauma is so ubiquitous in this population, a separate chapter addresses its influence on the cooccurrence of aggressive behavior.

In Chapter 6, research regarding the overreliance on medications to control the aggressive behavior of people with ASD and other IDD is presented. Factors that cause discomfort or pain provoke or increase the risk of emotional challenges and aggressive behavior as noted in Chapter 3. Many people with ASD or other IDD lack the means to reliably report their own illnesses. This is especially true in cases of discomfort or pain caused by the very medicines that were, ironically, aimed at reducing aggression. In fact, the most frequently used outcome measure in psychotropic medication studies for this population is the *Irritability* subscale of the aberrant behavior checklist (ABC). The subscale is predominantly composed of items related to various aggressive behaviors, noncompliance, and agitated outbursts. Concerns about polypharmacy and the long-term use of psychotropics to manage aggressive behavior are discussed. Commonly missed side effects associated with increased occurrences of aggression are also presented. Multiple case examples are provided, as well as tips for discovering side effects that are often overlooked, such as akathisia, orthostasis, constipation, urinary retention, and dysphagia. Research addressing resistance to tapering psychotropics with unclear efficacy is reviewed. Strategies for changing the pattern of searching for "the magic pill" are also addressed (Charlot et al., 2020).

In Chapter 7, the specific ways in which a history of trauma and early life stress may lead to a pattern of responding to stress with aggressive behavior are discussed. The chapter also examines why PTSD is infrequently diagnosed, even when reports suggest that a large percentage of people with ASD or other IDDs are exposed to trauma. Tips for detecting when aggressive behavior is likely provoked by past trauma experiences are provided, with multiple case examples. The need for caregivers to understand that traumatized individuals need to feel safe, cared for, and worthy of care is discussed as well. Links between trauma and aggressive behavior are carefully explored. Ways to provide trauma-informed care to people with ASD and other IDDs are presented, with detailed suggestions to improve the quality of life for these individuals (Harvey, 2016).

In Chapter 8, Child rearing is described as both difficult and rewarding. People with ASD and other IDD can present additional challenges and rewards as well. Experiencing physical aggression and the concomitant stress response can be debilitating and may interfere with a caregiver's resources, impacting both family members and paid caregivers. This chapter examines a structure to understand stress responses of this kind, along with strategies to mitigate their impact (Kwong and Low, 2025).

In Chapter 9, the approach described builds on existing ABA (applied behavioral analysis) models, especially Positive Behavior Support, and focuses on examining factors that influence the probability of occurrences of symptoms and behaviors, with utility for assessing and creating plans to address aggressive behaviors. The approach provides a means of incorporating multiple domains of influence on aggression into a single treatment plan. The developmental contextual assessment and treatment approach shifts the usual *Functional Behavioral Assessment and planning* focus from consequent events to one centered on ***contexts.*** This has been urged before by advocates for a more positive behavioral planning methodology, first described by experts like Ted Carr and Robert Horner (Carr and Horner, 1991). Critical contexts are those within which the occurrence of various behaviors and signs of distress is more probable. Aspects of heightened concern with contexts have been proposed previously and are amplified in the use of the suggested approach. Contextual factors may also be described as "influences." When a factor or influence is associated with a higher probability of the occurrence of a specific behavior or symptom, we can try to modify its impact proactively. Another difference between a contextual assessment and usual *functional assessment* is the importance, in the former, of discovering the contexts in which *happiness* is more probable. Many plans focus almost exclusively on risks for aggression and other negative targets. The developmental contextual assessment and treatment tools shared can be used to create a plan using multiple modalities. There is an emphasis on teaching and reinforcing positive "replacement behaviors," an area receiving more attention, though often misunderstood.

The premise of focusing on promoting an individual's *happiness* is based on the understanding that when people are happy and reasonably well, they tend to be less violent. By definition, increasing a person's happiness and wellness addresses challenging behaviors, including aggression. The data collection system used to perform a contextual assessment is described in detail. Case examples show how these assessments can generate broader case formulations and assist in addressing influences such as illness, side effects, sleep challenges, problems with a lack of meaningful engagement, difficulties that arise when with others with ASD/IDD, educational or programmatic problems, stress with transitions, difficulties with social-emotional understanding, and problems related to functional communication difficulties (Beasley, 2024; Hassiotis and Vegh, 2024).

Finally, in Chapter 10, we include our summary of the book's key points and perspective. We argue that a multidisciplinary and multimodality approach to understanding aggression displayed by people with ASD and other IDD is needed to develop effective treatment. This includes components described in the chapters above. In fact, to reduce its occurrence, one needs to focus less on managing it and more on uncovering the root causes of distress and frustration. The benefit of seeing wellness and happiness as treatment outcome goals for individuals who come to clinical attention primarily due to aggressive behavior is recommended. The hypothesis put forth is that aggressive behavior can be *contained* using traditional methods, but often at the cost of quality of life and frequently with repeated relapses. The approach suggested here aims to provide more opportunities to achieve a positive quality of life while reducing aggressive behavior. Suggestions for teaching caregivers are also provided.

References

Beasley, J.B., 2024. Addressing mental health equity for people with intellectual and developmental disabilities: recommendations for national goals. In: Kasey, R., Caoili, A., Beasley, J.B., Urquilla, M.P. Eds., Mental Health in IDD Workgroup Publications, University of New Hampshire, Institute on Disability, Durham.

Carr, E.G., Horner, R.H., 1991. Social influences on severe behavior problems in people with developmental disabilities: implications for applied behavior analysis. J. Appl. Behav. Anal. 24 (4), 437–459.

Charlot, L., Benson, E., Tasse, M., Fox, S., Hassiotis, 2017. Depressive Disorders. In: Fletcher, R., Barnhill, J. (Eds.), Diagnostic Manual – Intellectual Disability-II (DM-ID 2): A Textbook of Diagnosis of Mental Disorders in Persons with Intellectual Disability. NADD Press.

Charlot, L.R., Doerfler, L.A., McLaren, J.L., 2020. Psychotropic medications use and side effects of individuals with intellectual and developmental disabilities. J. Intellect. Disabil. Res. 0964-2633. 64 (11), 852–863. doi:10.1111/jir.12777.

Charlot, L.R., Doucette, A.D., Mezzacappa, E., 1993. Affective symptoms of institutionalized adults with mental retardation. Am. J. Ment. Retard. 98 (3), 408–416.

Charlot, L.R., Hodge, S.M., Holland, A.L., Frazier, J.A., 2022. Psychiatric diagnostic dilemmas among people with intellectual and developmental disabilities. J. Intellect. Disabil. Res. 0964-2633. 66 (10), 805–816. doi:10.1111/jir.12972.

Charlot, L. (2016), Multidisciplinary Assessment of individuals with Intellectual and Developmental Disabilities, Rubin, I.L., Merrick, J., Greydanus, D.E., Patel, D.R. Eds., Rubin and Crocker 3rd Edition: Health Care for People with Intellectual and Developmental Disabilities across the Lifespan, Springer, Dordrecht.

Charlot, L., Sheehan, R., Hassiotis, A., 2019. Use of Medications in the Treatment of Aggressive Behavior. In: Handbook of Intellectual Disabilities. Springer, Cham, pp. 661–681.

Harvey, K., 2016. Trauma-informed care [Article]. Frontline Initiat. 13(1), 1-12.

Hassiotis, A., Vegh, B., 2024. New Insights on Challenging Behavior: assessment and Treatment Across the Lifespan. In: J. L. Matson (Ed.), Handbook of Psychopathology in Intellectual Disability: Research, Practice, and Policy, p. 159–185.

Kalb, L.G., Stuart, E.A., Riley, A., Newschaffer, C.J., Vasa, R.A., 2012. Psychiatric-related emergency department visits among children with an autism spectrum disorder. Pediatrics 129 (6), 1061–1068. doi:10.1542/peds.2011-2306.

Kwong, A.N.L., Low, L.P.L., 2025. Meaning-Making Experiences of Family Caregivers Caring for a Member With Intellectual Disability and Challenging Behaviour. J. Policy Pract. Intellect. Disabil. 1741-1122. 22 (3). doi:10.1111/jppi.70014.

Matson, J.L., Adams, H.L., 2014. Characteristics of aggression among persons with autism spectrum disorders. Res. Autism Spectr. Disord. 8(11), 1578–1584 doi:10.1016/j.rasd.2014.08.004.

Rubin, I.L., Greydanus, D.E., Merrick, J., Patel, D.R., 2016. Health Care for People with Intellectual and Developmental Disabilities across the Lifespan. Springer International Publishing, United States, pp. 1–2307. doi:10.1007/978-3-319-18096-0.

Tassé, M.J., Havercamp, S.M., Krahn, G., Shogren, K.A., Bonardi, A., Kim, M., Witwer, A., 2024. About whom are we talking when we use intellectual and developmental disabilities? JAMA Pediatr. *179* (1),83-89.

CHAPTER 2

Developmental effects on aggressive behavior

Lauren Charlot
UMass Chan Medical School, Worcester, MA, United States

Introduction

The terms "developmental" and "neurodevelopmental" are both used in this chapter. In addition to autism spectrum disorder (ASD), as noted in the introductory chapter, the assessment and treatment approaches described here are ones designed for people with intellectual disorders or those who have other significant developmental challenges that require them to receive supervision or staff support. In almost all instances, these individuals have a neurodevelopmental disorder. This is because their developmental profiles include factors that impact neurological functions. These include atypical development with challenges in communication, social cognition, information processing, and planning, all of which are closely involved in the risk for aggressive behaviors. In Table 2.1, an overview of some of the more common features associated with neurodevelopmental disorders, including ASD and other intellectual and developmental disabilities (IDD), is provided. These features include deficits that alter the probability of aggressive behavior. A neurodevelopmental profile is derived from various psychological assessments and describes both one's strengths and weaknesses across several domains, overall describing how we think and interact with our environments. Developmental histories may reveal the timing of the development of typical abilities, such as physical growth, walking, talking, processing basic information, and relating to others. Over time, testing can be used to determine roughly how an individual's executive functions operate—something usually seen as a neurodevelopmental feature. The developmental capacities described in this chapter are primarily those known to be secondary to a neurodevelopmental disorder. In many instances regarding the people addressed in this book, older youth and adults may process information in ways more similar to those used by younger people with a neurotypical developmental history. Some aspects of developmental progress, such as the emergence of specific motor

Aggressive Behavior of People with Autism and other Intellectual and Developmental Disabilities
https://doi.org/10.1016/B978-0-443-26552-5.00006-0

Table 2.1 Overview of developmental and neurodevelopmental features seen in ASD/IDD.

Neuro developmental features	Description	Impact on risk for aggressive behavior
Theory of mind Competing perspectives	People may appear as if they are being selfish or unsympathetic when, cognitively, they do not comprehend that others may have a completely different perspective of the same phenomenon. Individuals with these challenges may only discuss their interests, failing to realize that something is usually said or done in social situations. Conversations lack reciprocity.	Caregivers and clinicians may view the person as oppositional, defiant, antisocial, or narcissistic when they are simply unaware of the impact of their words or actions on others. This can result in frustration, as people treat the individual as if intentionally inconsiderate.
Sensory sensitivities	Many people with ASD/IDD have atypical sensory profiles, with significant sensitivity to things like loud noises, sirens, visually over stimulating environments, light touch, or people getting too close, how clothing or tags feel on the skin, and even certain scents.	The individual may become aggressive to escape being brought into spaces or being engaged in ways they find painfully adverse, secondary to sensory phenomena; for example, a person with sound sensitivities placed in a residence with loud housemates.
Understanding social norms or social cognition	Not knowing what to say or what behavior is expected of you in interactions with others, while lacking an understanding that other people often have different perspectives, as mentioned above, contributes to frustration and stress in many social activities. Individuals may be rejected or may themselves avoid social events.	This can cause caregivers and others around the individual to become frustrated with them, sometimes contributing to social isolation. These difficulties can increase the risk of aggressive behavior.
Difficulty expressing—reporting internal states	People may be able to speak well and describe concrete phenomena or even some abstract ideas, but they struggle to report their own feelings and accurately self-report sensations that indicate an illness or injury.	When caregivers or clinicians fail to understand the true nature of a person's feelings or discomfort, these can become setting events that provoke or increase the probability of aggressive behavior.

Table 2.1 Overview of developmental and neurodevelopmental features seen in ASD/IDD—cont'd

Neuro developmental features	Description	Impact on risk for aggressive behavior
General difficulty in functional communication	People may use words and speak without using language to request things or explain their experiences to others, so their communication is not *functional*.	When caregivers mistake speaking for understanding, problems may occur, leading to an aggressive act.
Repetitive behaviors	Stereotypies such as stimming (waving fingers in front of one's eyes), hand-flapping, rocking, repeatedly lining up items, repeating phrases (echolalia), and insistence on sameness—i.e., even a route taken to a frequented spot must be taken every time to prevent severe anxiety.	Some repetitive behaviors help to contain anxiety and can even be self-soothing. If seen as abnormal or disruptive, efforts to stop them can provoke aggressive behavior.
Restrictive and overly focused interests	Intense fascinations with trains, cars, weather, digital media, video games, animals, numbers and patterns, music, etc.—at times, a person will only engage in discussions or activities about the topic on which they have become fixated.	If the drive to engage with and be involved in the area of focus is disrupted or prevented, aggressive behavior can result.
Assessment shows great difficulty with "set-shifting"	A person may struggle to move off a topic or process despite new information being provided that should prompt an adjustment. The *Wisconsin Card Sorting* Test assesses shifting capacity (see Zelazo et al., 1996).	Difficulty shifting (a frontal lobe function or executive function) often leads to *perseveration*. When caregivers try to move a person along or get them to shift, it can provoke an aggressive response likely related to anxiety.

disturbances and other medical conditions, can be tied to a person's genetics. Common syndromes associated with ASD and other IDD are described in Chapter 4. Understanding human behavior is complex, especially when trying to explain phenomena like aggressive behavior, whether or not one has a neurodevelopmental disability. It is also important to understand that experiences impact development, and development affects the experiences a person has. The field of developmental psychopathology focuses on this relationship and helps with understanding the transactional nature of the emergence of problems such as aggressive behavior (see Sameroff and Chandler, 1975).

Understanding outcomes related to neurodevelopmental disorders

About 75 years ago, neuropsychological views about what happens when a brain is lesioned began to change in significant ways that continue to be a cornerstone of assessment in this field. A.R. Luria is known for introducing a "systems approach" for describing mental functions that "assumes that one and the same mental function may be disturbed in the case of lesions of several different brain areas" (Luria, 1966; Mikadze, 2014). It is now well understood that how a patient with brain impairments behaves following an occurrence causing a lesion with subsequent impairment, or even impairments resulting from a *congenital* insult, cannot be simply localized to the site of the impaired brain tissue. Before imaging became an available diagnostic tool, clinicians performed assessments to try to determine the location of lesions. Luria pointed out the flaws in assuming a 1:1 relationship between a locus in the brain that is injured and the resulting deficits or even spared capacities that evolve over time post-injury. The process is dynamic, and the brain tries to compensate for deficits. The result may be tied to the extent to which the brain can take advantage of neural plasticity to relocate lost functions.

What clinicians see when assessing individuals with ASD/IDD may also be derived from multiple pathways. This is a concept often used in medicine. A fever, for example, is a *final common pathway* for multiple problems and conditions. The assessment that can point to an effective treatment for a fever must seek to identify what caused the fever we see today. This means the current one. We recognize it may not be the same cause as what led to a fever you had when you sought treatment one year ago. There are things we can do to reduce a fever even before we identify the cause of the current problem. However, even if we lower the fever and the patient starts to feel better, we may not have solved the problem. If the fever is due to factors that will cause illness to evolve and worsen, we really must try to identify them. Skipping the assessment of potential factors contributing to this clinical picture is not advised. These basic principles are helpful when trying to identify factors or sources for the aggressive behavior of a person with autism or other IDD. The goal is to find out why the patient is aggressive to direct treatment. We can sometimes use a treatment to "lower the fever," in this case the rate or intensity of aggressive behavior before we understand why it is happening now. In the case of many individuals with ASD/IDD presenting for psychiatric help with aggressive behavior, antipsychotic medications are the "fever treatment" that may work initially without solving the problem. Here, the focus is on those individuals who were medically treated for aggressive behavior and the treatments ended up less effective, leading to more medications being added while the aggressive behavior still occurs.

In these cases, it makes more sense to believe that the aggressive behavior acts as a final common pathway for distress and frustration experienced by individuals with few ways to demonstrate it (people with neurodevelopmental features associated with difficulty in self-reporting symptoms). Using this view, it is understood that short-term aggression-lowering interventions may, at times, only work in the short term. The tendency to demonstrate upsets behaviorally often has roots in the individual's neurodevelopmental profile. Assessment recommendations in this book are aimed at helping clinicians identify and understand various influences on the occurrence of aggressive behavior, including those related to the person's individual neurodevelopmental challenges. As suggested in Luria's work, the identified clinical problem may have complex roots and varied pathways that result in what we see at assessment. How individuals with neurodevelopmental challenges manage their world or cope with stress will differ from person to person, even with similar deficits, as well as at different times and in different contexts.

How neurodevelopmental features impact aggressive behaviors

Understanding an individual's unique manner of perceiving the world, processing information, and problem-solving helps us identify factors that impact the occurrence of aggressive behavior (see Table 2.2). For example, people who tend to "think in pictures" may become frustrated when all requests made of them are explained verbally. Some individuals struggle

Table 2.2 Traditional IQ or psychological testing—FSIQ is less helpful than other aspects of these instruments.

Atypical findings	Suggested interpretations and approaches to avoid aggressive responses
Pattern of one standard deviation or > between verbal and performance IQ may be best predicted early on by VIQ	Adjust how you teach and interpret the causes of aggressive acts, taking into account the variability and atypical nature of the findings
PIQ strengths and strengths in visual-spatial information processing may be occurring while verbal abilities are weaker.	Use visuals to teach and shape new skills, i.e., functional communication training (FCT), building daily schedules, and multimodal communication tactics
VIQ is a strength—strengths in the use of words and verbal processing	Augment teaching with lists, paired with words
Significant variability across subtests	Use stronger capacities to build more challenged ones, i.e., visuospatial strength—use maps, figures, pictures, and verbal subtests which are stronger than others; use lists with step-by-step tactics clearly spelled out
Working memory much lower than the scores for FSIQ, PIQ, and VIQ	Provide more structured support—use the three Rs: rehearsals, reminders, repetition, and multimodal communication tactics
Processing speed is slower than expected, given FSIQ, PIQ, and VIQ	Slow your communication, give time for a response, and ask the person to repeat back
If assessed or noted, ***set-shifting*** is more challenging than other cognitive processing	Use transition tools (an item to carry from one spot to another, a fun thing awaiting, something in hand distracting from hitting, etc.; reminders, first-then cards) Transition tools can help make shifting less stressful Assist with transitions using links across things to be bridged Think of ways to "bump the needle out of the groove"
If tested or noted, challenges to understanding competing perspectives, also known as theory of mind, and difficulty understanding social cues arise, yet other cognitive capacities are not impaired or are only minimally challenged	Recognize the significant difference between purposeful negative acts and simply misunderstanding the impact of one's behavior or words on others (missing how others feel); his is a critical concept to teach caregivers

Table 2.2 Traditional IQ or psychological testing—FSIQ is less helpful than other aspects of these instruments—cont'd

Atypical findings	Suggested interpretations and approaches to avoid aggressive responses
Challenges noted to social-emotional and intellectual skills	Teach social rules or expectations and shape their use in a concrete manner, i.e., reinforce + success by meeting step-by-step social expectations
Motivational influences	Increasing the motivation to improve skills and abilities can enhance outcomes. Use preferred stimuli to increase interest and apply positive reinforcement to shape successful trials during skill-building—if the person loves *Harry Potter*, incorporate references to it in the skill-building trials

with understanding social context and act in ways that prevent others from including them in social events. Subsequent disappointment and frustration can contribute to aggressive acts. A case example illustrates this concern:

Kevin, a 19-year-old man with ASD, seemed to prefer spending time alone in his room rather than being with his housemates. His treatment team decided it was important for him to learn to tolerate group activities. He then began to "tease" his housemates and even escalated to assaulting them, or they would assault him. Following these incidents, he would be sent back to his room. His team decided he "liked" teasing his housemates. However, it gradually became clear that he was anxious during these programmed social activities, and he learned he could "escape" them quickly by engaging in provocative acts. Although he could speak well, he faced challenges in social awareness and did not express himself effectively using words. It was also recognized that features of his autism included sensory sensitivities to sound and difficulty with others getting too close to him. These neurodevelopmental autism-associated features were important to understand when assessing why he became aggressive in the context of group activities at home. Kevin learned best when he was gradually exposed to things that caused him anxiety, such as "too many people," loud environments, and people being "too close" to his personal space. An alternative plan was developed to help him enjoy one-on-one time with a person whose company, it was discovered, did not make him anxious as a first step. His caregivers learned that he was not aggressive simply to escape things he needed to do, but rather as a reaction to the stress caused by the push for him to be with multiple housemates.

To identify factors that add risk for aggression, it is important to know how a person functions when well and at their "best." As suggested, people react to stress in ways tied to their specific neurodevelopmental capacities and challenges. Young children with a neurotypical developmental history may become quite agitated when anxious (e.g., have tantrums). This is less common for older neurotypical individuals. Adults with autism and other IDD may respond to stress more similarly to how a younger neurotypical person would, in part due to similarities in how information is processed or social norms are understood. Some people find it difficult to shift responses and may be described as having problems with flexible thinking or "set shifting." This may occur among people with ASD but also other IDD (e.g., exposure to substances in utero) more so than it does with individuals who have a neurotypical developmental history (Rangmar et al., 2015).

People will respond best to interventions aimed at reducing aggressive acts when these interventions are informed by an understanding of their profile of strengths and weaknesses in the capacities needed to learn and retain information. This includes the person's general means of information processing and planning, as well as their expressive-receptive verbal abilities, visuospatial processing, and social-emotional intelligence. Other related neurodevelopmental challenges, such as sensory sensitivities, also play a role in the generation of aggressive acts. The ways in which the assessment of aggressive behavior can be enhanced by including information found in standard cognitive testing are briefly reviewed below. This can serve as a starting point if broader-based evaluations, including measures of executive function (neuropsychological testing), are unavailable. A quick review of common neurodevelopmental features of autism and conditions associated with intellectual disability is provided in Table 2.1 to aid in creating developmentally sensitive interventions (Fitzpatrick et al., 2016).

Variability in cognitive profiles, learning, and aggressive behavior

In general, it is widely understood that people with a more restricted repertoire of possible responses to stress are more likely to react "behaviorally." For example, children who have a profound or severe ID diagnosis are more likely to express frustration or distress in the form of aggressive acts than are youth with a neurotypical developmental history. Other factors related to information processing and planning abilities may impact the range of possible or likely responses to stress. The most commonly employed tests of general intelligence are the Wechsler Adult Intelligence

Scale (WAIS), the Wechsler Intelligence Scale for Children (WISC), and the Stanford-Binet (SB). ADD citation here (Roid, 2003; Wechsler, 2014, 2017). These tests yield quite similar profiles of how people process information, testing subunits of cognitive capacities, as well as describing general intellectual ability. The overall measure of general intelligence is known as a *Full Scale IQ (FSIQ)*. In many systems of care, these assessments are paired with measures of adaptive functioning to determine whether a person qualifies for Developmental Disabilities services. As a result, people may be found to be eligible for specially defined and funded services. The addition of adaptive behavior assessments was a way of recognizing that FSIQ alone can be misleading and not always the best tool to inform plans addressing a person's developmentally driven service needs. There may be people with an FSIQ of 69 who need minimal added assistance and function well enough without having "special" services. Alternatively, some people with an FSIQ of 76 or 80 (even higher) may be so adaptively challenged that they need specialized assistance to be safe or just to get by day to day. For the purposes of understanding aggression, standard IQ tests are less helpful when people have more significant levels of challenges, such as when an FSIQ falls at 40 or lower. In these instances, adaptive behavior assessments may be more useful in considering a person's support needs. This is especially true when the individual's scores are at the "floor" of multiple subtests that are needed to compile a valid FSIQ estimate. See Table 2.2 for a brief overview of the uses of IQ testing to understand pathways and treatments for aggressive behavior.

For the more narrow and specific purposes of assessment and treatment of aggressive behavior, standard tests of intelligence can be used for different purposes. They offer a characterization of a person's general abilities for information processing and planning, and this is helpful when assessing pathways to aggressive behavior. In the case of IQ testing, it is even more helpful to know whether the person's *pattern* of information processing reflects areas of significant strengths and deficits in their developmental profile. Variabilities, when present, may help to predict or explain why aggressive acts occurred in certain contexts. A person might respond to an intervention aimed at preventing aggressive behavior better if their strengths are used to teach new, safer responses to stress. A standard IQ test might reveal which skill areas could potentially be enhanced with education or the shaping of new skills. Part of the process, too, is determining roughly how much work we must do to teach new skills that will help the individual be more autonomous. The capacity for processing information presented verbally versus information shared visually may differ enough to warrant a mention in testing reports, such as when there is a full standard deviation or more between these two domains. As noted above, people likely learn best when they start using their strengths in how they learn. We also want to enhance areas where more challenges are evident. In some instances, working memory and processing speed challenges will affect the risk for aggressive behavior. Some individuals express many things verbally in a manner superior to their ability to process or problem-solve incoming verbally presented information. This frequently results in caregivers overestimating the person's understanding in ways that lead to labels such as "oppositional" or "noncompliant." When a person doesn't "comply" because they are confused about what is being asked, the intervention is quite different from when a person wants to maintain control and fully understands the request or "demand."

If a person shows relative strength in visual processing and weakness in verbal arenas, visual materials should be used in the treatment plan. Even when this is known, at times, service providers don't use multiple modalities in daily communication. This seems to happen when a person with autism or other IDD speaks well. For example, during consultations, caregivers tell me, "This person understands, and there is no need to add visuals." In such situations, there can be a tendency to equate speaking with understanding. Yet, for some people with autism or IDD, the two abilities are not equal. Individuals with autism are known to learn "scripts" that fit various situations. A person may seem to understand the typical expectations and yet only know, "This is what people say," when doing a certain activity or when in a particular situation.

Case example: A 17-year-old learned to greet his aunt before dashing off to play video games when he came to family parties at her house. He learned to ask to "reciprocate," asking his aunt how she was after she asked him. Initially, it was very clear that he learned what to say but had no insight into why he was saying it, apart from his mother instructing him to do so. She did not lecture him or moralize about this. He loved his aunt quite a lot and would hug her spontaneously. He also noticed the positive reaction to his greetings and that his mother was pleased. His lack of reciprocity and understanding of the "social norms" for attending a party at his aunt's house was not a reflection of him being inconsiderate or antisocial. He simply did not know "the rules" or social expectations. He also had no idea it would make his aunt feel bad if he did not come and say hello when he arrived.

When he was younger at school, other students would "get too close." This would trigger a panic response, and he became aggressive. This, in turn, caused him to be labeled as conduct-disordered. One day, he was removed from the school in handcuffs. When he explains this now, it seems shocking because he is a sweet person. He recalled doing things he did not know were "wrong," like taking a full bowl of a favorite food off the buffet for himself at a gathering. He was bullied for telling the teacher when students snuck out of the room after she stepped out and said, "Everyone stay in your seat." He did not understand the concept of lying, even in high school. At the same time, he could troubleshoot computer software problems quite efficiently. With enormous hard work by his mother, he was moved into a new school where they immediately noticed he was a youth with autism. He ceased to have violence at school as they accommodated his needs for space and the use of a tablet to write. It seemed remarkable that his initially diagnosed psychiatric problems were resolved simply by people around him understanding his neurodevelopmentally caused difficulties.

When individuals who have an average or above-average FSIQ (Full Scale IQ) are still very challenged in understanding social cues, this needs to be incorporated into treatment plans. Plans crafted to enhance these weaker areas or capacities can be helpful in the prevention of aggressive and other behaviors that challenge. Plans that are proactive and not simply responsive or "reactive" can employ interventions to build, teach, or shape social awareness and social skills when these areas need to be enhanced. Doing so can prevent conflicts arising from viewing the person with challenges in social cognition as rude or conduct-disordered when they do not understand something because of the nature of their neurodevelopmental profile. In many instances, individuals may learn to ask for a break or to request help to get out of an anxiety-provoking situation. The further step needed is to work to understand what makes a given situation one in which a person will become aggressive to be rescued from facing it. Plans that focus on "compliance" sometimes miss this opportunity.

When caregivers speak to an individual without supplementing communication visually, many people will struggle to understand what they are being told or asked. This can happen even when a person speaks fluently. Resisting a request (or "demand") may stem more from a failure to truly understand it than from a desire to be uncooperative. In the reverse scenario, some individuals process information better when it is written or spoken rather than presented in other ways. In either case, the clinician writing a plan will benefit from knowing each individual's strengths and areas where they may need more support. In a number of instances, problems with information processing speed might mean an individual simply needs more time, along with more rehearsals or reminders, to cooperate with important requests. Too often, we inadvertently increase the risk of an aggressive act by failing to make this distinction.

In cases like those described above, aggressive behavior may be more likely to occur when caregivers expect that a person is intentionally being "noncompliant," but the individual actually does not fully understand what is being communicated or asked of them.

A multidisciplinary evaluation team assessed a 34-year-old gentleman with ASD, who was described as having mild ID. However, upon examination, it became very clear that he had significant working memory challenges and minimal social awareness. Despite this, he was typically a sweet, gentle person who experienced major outbursts of aggressive behavior intermittently. It was revealed that one of his worst outbursts occurred in a grocery store during a shopping trip. Subsequently, if something reminded him of shopping or if someone mentioned it, he would perseverate about going shopping or even fixate on the name of the store. One of the clinicians hypothesized that he did this intentionally, knowing it would intimidate staff, as his repetitious statements often preceded aggressive behavior. However, it was evident that he was intellectually or cognitively unlikely to even be capable of such calculated behavior. More likely, he now associated the store with the event that took place there, an event that he understood had caused much negative feedback and distress for everyone involved. To better address this, it was recommended to determine whether all mentions of shopping were a trigger or only when mentioned at a certain time of day, by certain people, or in specific contexts, such as when the environment was loud or crowded. It was discovered that he was often very foggy and still fatigued when he woke in the morning, and this was consistently a time when triggers might provoke aggressive behavior. Approaching him by allowing more time before making requests and adjusting the timing of his evening medications helped to resolve his early morning dysphoric mood.

In almost any case, "noncompliance" with a "demand" may not be an issue if we are careful in how we make *requests*. When something we know the person needs to do is actually framed and delivered as a *request,* the person may decline to cooperate but still retain some sense of agency.

There are a number of ways to reduce a negative response to requests. In many cases, the person's failure to follow directives is seen as being caused by a history of reinforcement driven by motivation for attention or escape. The interventions, at times, are then designed to prevent escape or reduce attention. This often fails for several reasons. One significant cause of this interpretation and approach failing is that the person did not *process* the request. Failing to process a request may, in turn, be caused by issues with working memory. Furthermore, it is important to know that people can even repeat back something said to them without truly understanding or gaining *insight*. Processing requests when tired, hungry, or simply not feeling well may increase the likelihood of a problematic response. Starting with a more comprehensive understanding of the individual's specific cognitive skills and challenges can be helpful and can speed up a clinician's ability to identify tactics more likely to be effective, beginning with a person's strengths. In doing this, it helps to understand how memory works. IQ tests provide some insight into this when there is a score reflecting working memory abilities. We find that some people with IDD can follow only one or two steps at a time. Knowing this is critical to helping a person succeed.

Caregivers may need training to understand that speaking does not equal understanding, as they may believe the person must understand when he or she is repeating "scripts" or statements they have learned are the "right things to say" under various circumstances. They may learn what to say without knowing why they are saying it.

The discrepancy between verbal intelligence IQ (VIQ) and performance IQ (PIQ) found in research regarding the cognitive abilities of people with ASD is associated with ASD symptoms (Johnson et al., 2021). It has been suggested that VIQ is the "critical predictor." Given that VIQ is a component of the VIQ-PIQ discrepancy, it was unclear which factor was most influential in a study where data from 527 children and adults with ASD, sourced from the *Autism Brain Imaging Data Exchange* (ABIDE), were analyzed. Johnson and colleagues examined the contribution of VIQ versus PIQ to the noted discrepancies and explored the relationship to ASD core symptoms. "VIQ was most associated with lower symptom severity as measured by the Autism Diagnostic Observation Schedule (ADOS) total score... when VIQ was included in models predicting ASD symptoms, associations with PIQ and IQ discrepancy were not significant." This suggests that previous findings of an influence from PIQ on these overall gaps between the two scores were most likely mediated by the connection with VIQ. Their analysis concluded that an association between VIQ and ASD communication symptoms correlated with the severity of ASD symptoms. "Subtyping ASD on the basis of VIQ should be a point of future research, as it may allow for the development of more personalized approaches to intervention." However, it is critical to determine if VIQ is relatively stable across development. Longitudinal studies show that many features initially impaired will improve over the course of development. However, while spatial and "manipulation" skills develop positively for youth with ASD, "language does not develop." Positive outcomes are predicted by baseline levels of language capacities. In addition, in general, the more "autism-specific" symptoms young children have, the greater their risk of a poor cognitive outcome. The observed heterogeneity at the individual level "is not random and is a true marker of the development of children with ASD." This heterogeneity was stable across multiple countries of origin examined (Mauga et al., 2016). Both differences and similarities between individuals and groups with neurodevelopmental disorders "might be explained by more specific variables, for example, the genetic, biological, cerebral, and developmental trajectory characterizing each child with ASD" (Bernard et al., 2019).

How executive function challenges underlie mood dysregulation and problems with self-control

Some individuals with IDD have major difficulty executing neurotypical cognitive and social tasks, the adequate execution of which depends on the functioning of the frontal lobes. Frontal lobe-mediated "executive functions" noted above are important for reasoning, processing information, problem-solving, and planning, as well as using experiences from the past to inform future behavior. Many people with IDD will eventually learn things, but often not without a lot of support and repetition. Because preventing influences on aggressive behavior can be labor-intensive, we are drawn to interventions that start after the fact or are *reactive*. Caregivers also want there to be fast relief from the struggles. Generally, it is not so easy or simple.

One place to start is to assess the extent to which a lack of control over a person's own mood and behavior is initially caused by or amplified by executive function deficits. These are not always tied to overall IQ. "Regulatory processes collectively termed executive function (EF)" are more predictive of school success than IQ and are also predictive of adult life outcomes. Broadway et al. (2010) summarized EF as having been empirically demonstrated to encompass "cognitive control ... reflecting prefrontal cortex-mediated processes such as working memory, response inhibition, shifting, and response selection … in addition to planning." See Table 2.3 for a brief, simple overview of executive functions.

There is some evidence that working memory (WM) itself may be tied to flexibility of thought and that, across various diagnostic groups, youth with ASD have more difficulties in this domain. Further, WM may play a role in the occurrence of anxiety, which, as noted, is a factor that can drive aggressive responses (Sanders et al., 2017). Altogether, clinical focus may be enhanced by paying more attention to measures of EF and patterns within this area when developing plans aimed at reducing aggressive behavior and improving quality of life (Broadway et al., 2010). WM difficulties may play a role in the susceptibility of youth with ASD to trauma and to more severe reactions to traumatic events in their lives (Rumball et al., 2021; Schuiringa et al., 2017). As noted above, longitudinal investigations are important when any of the EF components or ASD/IDD features may change during development (Lord et al., 2015). Executive function and theory of mind (ToM) can show improvements over the course of adolescence. However, relative difficulty was still evident in contrast to subjects with a typical neurodevelopmental history (Roselló et al., 2021).

Working memory is likely a key component in the capacity for self-regulation, including mood regulation (Broadway et al., 2010; van Nieuwenhuijzen et al, 2021) As such, when impaired or diminished, it may influence the occurrence of aggressive behavior. It can be useful to consider aspects of working memory control (WMC), including attention and retrieval components. Learning a new behavior, especially one to replace the function of a current one that is dangerous, means attending to the lesson. It is interesting that we can repeat back the thing we were told immediately after, yet may forget it soon thereafter.

Table 2.3 Executive function.

Working memory	Hold the input in your mind for quick use
Planning	Create steps that you will follow to achieve your goal
Mental flexibility	Shift gears when a new response is needed, or when there is new information to consider
Inhibition	Stop and think before acting
Generativity	Generate novel ideas and behaviors spontaneously
Self-monitoring	Look for errors and change your responses based on experiences

Most people can recall a time when they forgot something very simple, something they know well. Usually, in these cases, there is a reason that we do not easily "retrieve" the information, plan, or respond when a response is needed. To remember something, like how to respond to certain requests, one must first retain the information and "store" it in the brain. We may be taught a response and repeat it back in a condition where we are calm and have support, only to later forget what we "learned." Often, this is improved through repetition. After a while, it becomes easier to remember what we learned. If we do learn a response and have stored it, to use it, we must be able to retrieve it. We may need to access that stored, learned response under conditions quite different from the ones in which we learned it. So, how is information stored in our brains? What factors affect recall?

Generally, we store parts of information by associating them with closely related data already stored in our brain. Like a library, the art section is separate from the geography section so we can more easily find the book we are seeking. Various stimuli or cues help us retrieve the information we have stored when we need it. The more "strengthened" neurological connections that have been made, the stronger the memory will be (again, usually with repetition). So, we practice and learn things if the material we need to learn is within our capability. People with IDD may be able to follow two steps in a task but have too much difficulty when there are three parts. Examining past assessments can help us better understand how to teach skills, but also ensure we do not expect too much from individuals and become upset with them when they fail to meet our expectations. Or, as suggested here, we want to avoid the assumption that not cooperating is always about refusing to do what we are told. As suggested, it is better still that we ask and make requests rather than present task "demands."

It is also important to remember that anyone can learn a skill or response and then be unable to retrieve it when anxious, uncomfortable, distracted, or ill. The setting—environmentally, but also the social-emotional context—may impact a person's likelihood of cooperatively responding, even when the individual knows how to do what was requested.

How working memory is assessed may impact findings. Children with ASD have been noted to perform better on visual-based than auditory-based working memory tasks. Furthermore, intellectual functioning was "correlated negatively with ASD symptom severity but positively with adaptive communication skills" in one investigation (Audras-Torrent et al., 2021). Overall, as noted above, FSIQ may be much less of a concern when looking for pathways to aggressive behavior than understanding how people best take in and use information, including what kind of external support and what kind of skill-building is likely to be helpful. Studies demonstrate that aspects of EF can be enhanced and show improvement with specific supports and teaching trials.

Mental flexibility and set shifting

Inflexibility of thought, the difficulty with shifting set even when new information is at hand suggesting a need to change, is a common source of distress in the lives of many with ASD and other IDD. For people who lack cognitive flexibility capacities, unplanned changes may provoke anxiety that, in turn, drives an aggressive act. The need to incorporate new information to inform a decision or plan of action can be highly stressful for some individuals with ASD or other IDD. This is complicated by the frequent fact of people receiving services within which they may be given little choice or control, adding to the frustration that can also lower the threshold for an aggressive outburst. There are tests for assessing mental flexibility, such as the Wisconsin card sort (see Table 2.1). A lack of the ability to set-shift is the source of "perseverations" (Kopp et al., 2023).

The ability of children with a neurotypical neurodevelopmental history to switch gears mentally on a Wisconsin card sort-type task is impacted by developmental capacities. In one study, the cards showed toys with different shapes. Most 3-year-olds tested continued to use the "pre-switch rules" during the "post-switch phase." In other words, the children were told to sort items based on color, and once proficient at this, were reinstructed to sort by shape. The children could state back to the examiner that there was a *new rule* now, and even state what that new rule was, yet they persisted in applying the first rule. These findings indicate that knowing rules is sometimes insufficient to permit their use. We sometimes see this effect in children who are older or even in adults when they have a neurodevelopmental disorder. Having difficulty adjusting to even small changes is a hallmark challenge seen in youth and adults with ASD and other IDD. These challenges seem to provoke frustration and, at times, aggressive behavior. We often speak of "perseveration" as a challenge for many people in this population. Perseveration may be a manifestation of a brain-based difficulty in "set shifting." We sometimes tell caregivers that a person may need our help to "bump the needle out of the groove." When these kinds of challenges are neurodevelopmentally driven but treated as purposeful acts of noncompliance, individuals continue to struggle, become frustrated, and this can provoke an aggressive act or worsen existing aggressive behaviors. When we recognize this for what it is, we can help reduce the stressful nature of the process and thereby avoid escalations to aggressive behavior.

A number of interventions may help people who get "stuck" and cannot seem to transition using new information. For some people, distraction may work better than preparation for change. See Box 2.1 for suggestions to prevent and improve outcomes when there is transitioning anxiety.

BOX 2.1 Transition tool kit

Although transitions can be challenging for people on the autism spectrum and others with IDD, there are strategies that can help make them smoother, including the following.

- When doing something the person likes a lot and has difficulty ending, explain what will happen at the end. Choose a highly valued thing to happen after ending the preferred activity. Use a FIRST-THEN approach. First, you shop, and then when we get home, you are having your favorite snack—xxxxxxx, or we will be playing yyyyy, you and I, and it's going to be fun. Then you can say, "Time to get to that game I promised you—so let's get going to the car."
- We make a lot of choices throughout the day, especially when we are faced with difficult or unpleasant tasks. We need to give our clients as many choices as possible to ease transitions. "We need to go… do you want me to drive by the park or by the water?"
- "Can you help me bring this back, and then did you want to play Uno or Checkers?"
- Sandwich harder activities between two preferred activities, and consider using visual schedules and timers.
- Have something reinforcing, fun, and desirable handy.
- Just word choice can make a difference, i.e., avoid the word "work."
- Spend a few minutes at the beginning of each task to be done—pair yourself and the materials with fun things to occur when it's time to transition. If a person is not moving toward the next activity without problems, the demands are probably too high, or the rewarding activity or thing to follow is not enjoyable or desirable enough.
- Give advance notice before a transition is about to occur.
- Announcing that it's time to switch gears without providing any notice can be quite distressing for a person on the spectrum. Advance notice helps the brain get ready to shift to another task or activity, thereby relieving anxiety.
- Time can be an abstract concept, so when referring to time to help with a transition, it is often best to represent it visually. Timers help

BOX 2.1 Transition tool kit—cont'd

provide a concrete reminder—there are some great ones that use colors to show a portion of time remaining. You can use a simple egg timer or any number of higher-tech visual timers, such as visual clocks on tablet or phone apps.

- Alternatively, you can avoid references to time by showing a visual representation of what's coming next. Such supports can be delivered through a visually supported schedule or by presenting drawings or photographs of the upcoming task or event.
- Timers can help, too, with waiting for a happy thing when it's difficult to wait.
- Reminders leading up to the change can be helpful, and you can even use a card with something fun pictured on it and say, "Here you go—10 more minutes..."
- Some people want to be doing something, and doing something becomes the focus instead of being worried or anxious about the transition. For example, when a person leaves, she takes an item to the next place or event and is thanked for helping. At times, keeping hands busy by holding something during transitions is helpful and reduces challenges with striking oneself or others. It's best if the person can bring something to the next place or event that is fun or is known to be associated with positive feelings.

Use structure and consistency

- Structure and consistency will help reduce the amount of work the brain needs to do to make a transition. It helps to keep materials for upcoming tasks or activity events in an easily identifiable and consistent place. Keeping the general order of daily tasks consistent can also help make transitions become more automatic.

Use simpler language

- When people are having difficulty with making transitions, there is a temptation to use language to try to move them along. This often backfires for people with ASD. More language can lead to verbal overload, which will only increase anxiety and make the transition take even longer.
- Instead, use a few keywords.

For example, consider the difference between the two sets of directions:

Example A:

"Come on. Get your socks and shoes on because we are going over to Auntie Mary's for your cousin's birthday party, and don't forget what happened last time when we were late—they started without us, and we didn't get any cake!"

Example B:

"Time for your cousin's party. Socks and shoes on," (parent holds up socks and shoes).

Continued

BOX 2.1 Transition tool kit—cont'd

Please note: you do not want or need to talk to the person like this all the time—only when you need to get a point across clearly. Stress will reduce the person's ability to use cognitive functions like verbal processing.

Provide "light praise" for good transitions.

Gently point out when the person has made a positive transition, and briefly explain *why* it is positive. This will encourage the person to be more receptive to working on making quicker transitions.

Key takeaways

- Transitions can be challenging for people on the autism spectrum.
- There are strategies that can help make transitions smoother, thereby relieving distress and frustration—the link refers to things outlined for children, but you can creatively adapt some of them for a young adult.
- See https://www.merakilane.com/visual-routine-chart-kids-autism-16-ideas-visual-learners/.
- Consider whether a usual transition can be planned to slow down a little. Any sense of rushing may trigger more anxiety.
- Social stories can be helpful when a frontal attack on a problem causes anxiety. Social stories provide something to rehearse so that a person is prepared once the situation described actually takes place, which can be an excellent strategy for helping people through transitions.
- See https://raisingchildren.net.au/autism/communicating-relationships/connecting/social-skills-for-teens-with-asd#social-stories-nav-title.
- Equipping a person with calming strategies can help him or her manage big emotions during transitions:
- PRACTICE DEEP BREATHING: It's no secret that taking deep breaths can help restore a sense of calm when anxiety hits, and a great way to teach this concept is to actually blow bubbles when she's feeling anxious. You can keep a stash of mini bubbles easily in a backpack, purse, or in the car to get started, and once the concept is mastered, have the individual practice blowing bubbles without the wand. This will give her an effective coping strategy she can use anywhere and anytime.
- USE concrete, positive, rewarding things to signify success: We all know we are motivated by the promise of a special outing or extra time to do a fun, preferred activity—and these can encourage a person through difficult transitions.
- CREATE A *CALMING* BOX: Another way to make transitions easier is to ensure the person has tools readily available when big emotions strike—fidgets, squeeze balls, or chewing gum.
- Add favorite activities to schedules or routines so transitions are, at least some of the time, about doing something relaxing (as suggested—you can sandwich these).

The combined impact of deficits in working memory, set shifting, and inhibiting responses on mood regulation and behavioral self-control is significant. It is most helpful to examine these areas rather than relying simply on an FSIQ to predict or understand aggressive behavior. Tactics to bolster working memory and set shifting can be used to lower the risk of aggressive behavior.

Competing perspectives, empathy, and aggressive behavior

Some people with ASD struggle to fully comprehend that other people perceive things differently, even when they have mild or no intellectual disability. It is important for clinical assessments to differentiate narcissism from the egocentricity of thought. Understanding that other people have thoughts separate from one's own is also known as *Theory of Mind*. In neurotypical development, very young children often have difficulty understanding that others see things from a perspective other than their own. Neurotypical children may begin to comprehend the concept of "competing perspectives" by about age 3–4 on average, though it is not a *binary* capacity. Gradually, the child not only recognizes that this occurs but also understands how it impacts their interactions with others. To engage in reciprocal interactions, an individual needs to be able to think about something that might interest the other person in the interaction. Otherwise, one's conversations tend to be "one-way," a socially problematic tendency. Some people with autism will speak to others only about their own interests—not because they are selfish, but because they simply do not think about what anyone else might be interested in. Interactions like this, where a child or adult with a neurodevelopmental disorder only addresses what they are interested in, can form the basis of conflict with peers who do not share this tendency. In many cases, the youth with a disability may be teased or, more frequently, avoided. Either consequence can contribute to the person with autism experiencing low mood, increasing the risk of aggression when frustrated or "triggered." Some youth and adults with other etiologies of a neurodevelopmental disorder, such as exposure in utero to substances, may show similar tendencies in terms of social awareness (Rasmussen et al., 2013; Zelazo et al., 1996). Children with ASD who describe themselves as wanting friends may not understand why they have difficulty making friends. Some avoid social interactions because these are associated with negative outcomes.

In the case of caregivers in situations like group residential care, problems arise when staff believe the resident is selfish or antisocial and treat them accordingly. In fact, educating caregivers can help reduce the added risk of aggression stemming from negative staff responses. Treatment needs to focus on helping the person with a limited ability to see competing perspectives learn how to interact in ways that enhance their social skills. In many cases, staff may intervene by telling the person that they are being mean and asking them to consider how the other person feels. This may be an unhelpful approach, even a provocative one, when the individual simply does not have the usual ability to imagine how others feel. In some ways, it would be like expecting a toddler to be more considerate of others. Again, the disconnect between some areas of strength and those with greater challenges can "fool" us into believing acts are purposeful when they are basically a manifestation tied to the individual's profile of cognitive capacities. It follows, then, that how we help individuals work toward a more positive outcome needs to be designed based on these influences.

What is "wrong" or what is impaired when a person acts aggressively toward another individual? A person with challenges in understanding competing perspectives doesn't realize that the effects of events may be quite different for another person than for themselves. This feature, sometimes described as a product of one lacking "Theory of Mind," is a phenomenon seen to varying degrees, but often in people with ASD. Baron-Cohen suggested that a person without a neurodevelopmental disorder who may have antisocial tendencies or a conduct disorder has intact cognition or understands what they are doing but seems to lack an accompanying affective component of feeling remorse about the action. Feeling bad for the person they targeted because of how they made that person feel is only possible when a person can conceptualize how others may feel. In the case of a youth with ASD who doesn't fully grasp that their action was hurtful, they also may not display remorse or concern. However, in this case, it is not because they understand and do not care, but because they lack understanding. The child with ASD cares but doesn't understand, and the conduct-disordered child understands but doesn't care. The site of the deficit is different, so the interventions that may help may be very different as well. A colleague described working with an individual with ASD and a moderate to mild ID diagnosis who assaulted her when he was frustrated trying to complete a task. Moments later, he was calm. He then asked her, "Will you play cards with me?" She described him as completely unaware that people you hit may not want to play a game with you right after you hit them. Other staff in the area saw him and thought his actions were purposefully done, that he was being "bad" and not caring. For others observing, he seemed to be a person who lacked empathy. He actually just did not connect what he did with the experience of the person to whom he did it. Again, the importance of knowing this is to create a remedy that fits. In his case, teaching him the connections between certain acts and outcomes, and a program aimed at increasing his capacity for social awareness, was needed rather than a "consequence" for his lack of apology.

One good question for clinicians to ask themselves is, "Is the intervention inadvertently designed to help caregivers feel better?"

One might be able to teach the script of an apology, but it might only be the caregiver who truly feels better after receiving it. The intervention must always be designed to improve the well-being of the individual we are treating. When treatments are effective, they also help caregivers. This is because when people are functioning better, they tend to engage in far less aggressive behavior. The goal needs to be to help the patient, resident, or family member function better in their environment, and gradually do so as independently as possible. Understanding the neurodevelopmental feature of having a concept of "competing perspectives" is required to interpret and design an intervention that can reduce aggressive acts exhibited by a person with deficits in this capacity.

A comprehensive assessment approach for understanding aggressive behavior

The work done to understand a person's history (their life story) and the context of their unique neurodevelopmental profile are the first critical steps toward developing a comprehensive, multimodality treatment plan for aggressive behavior. Modalities that may be used include many varied forms of intervention, such as those based on Occupational Therapy assessments of sensory sensitivities, music and art therapies, Speech and Language services that teach improved communication, and others. Once these steps are completed, it becomes much easier to understand the factors influencing the probability of occurrences of aggressive behavior and what tactics will be most effective when teaching skills to obviate reliance on aggressive behavior as a means of controlling one's world or expressing frustration and distress (Fitzpatrick et al., 2016). In the chapters that follow, ways to enhance assessment and build a more effective plan to treat people presenting with aggressive behavior will be described.

It is important that caregivers know that you can respect an adult as an adult while helping them in ways they can understand, based on your knowledge of their profile of neurodevelopmental strengths and challenges, which at times may resemble those we see in children with a neurotypical developmental history. If I need to repeat things and give multiple reminders, it is not being disrespectful of the person's adulthood. If a 35-year-old prefers watching kids' programs, I respect their choices and don't judge them as having something wrong just because they are different from most adults with neurotypical histories. Some adults without an IDD prefer children's movies and shows. The importance of understanding someone's unique manner of processing information, understanding relationships, and planning responses means you will be better able to help them. This includes knowing when they are behaving in ways we would never expect from them when they are well, or when they are simply having a more difficult time due to one of a variety of reasons associated with their unique ways of dealing with their lives. We need to identify both developmentally driven and contextual influences on people's behavior to provide effective help.

References

Audras-Torrent, L., Miniarikova, E., Couty, F., Dellapiazza, F., Berard, M., Michelon, C., ... Baghdadli, A., 2021. WISC-V profiles and their correlates in children with autism spectrum disorder without intellectual developmental disorder: report from the ELENA cohort. Autism Res. 14 (5), 997–1006.

Bernard Paulais, M.A., Mazetto, C., Thiébaut, E., Nassif, M.C., Costa Coelho De Souza, M.T., Stefani, A.P., ... Adrien, J.L., 2019. Heterogeneities in cognitive and socio-emotional development in children with autism spectrum disorder and severe intellectual disability as a comorbidity. Front. Psychiatry 10, 508.

Broadway, J.M., Redick, T.S., Engle, R.W., 2010. Working memory capacity: self-control is (in) the goal. Self Control Soc. Mind Brain 1, 163–174.

Fitzpatrick, S.E., Srivorakiat, L., Wink, L.K., Pedapati, E.V., Erickson, C.A., 2016. Aggression in autism spectrum disorder: presentation and treatment options. Neuropsychiatr. Dis. Treat. 12, 1525–1538. doi:10.2147/NDT.S84585.

Johnson, C.N., Ramphal, B., Koe, E., Raudales, A., Goldsmith, J., Margolis, A.E., 2021. Cognitive correlates of autism spectrum disorder symptoms. Autism Res. 14 (11), 2405–2411.

Kopp, B., Al-Hafez, B., Steinke, A., 2023. Habits, goals, and behavioral signs of cognitive perseveration on Wisconsin card-sorting tasks. Brain Sci. 13 (6), 919.

Lord, C., Bishop, S., Anderson, D., 2015. Developmental trajectories as autism phenotypes. Am. J. Med. Genet. C Semin. Med. Genet. 169 (2), 198–208.

Luria, A.R., 1966. The problem of localization of functions in the cerebral cortex. In: Higher Cortical Functions in Man. Springer US, Boston, MA, pp. 5–38.

Mauga, S., Café, C., Almeida, J., Marques, C., Duque, F., Oliveira, G., 2016. Intellectual profiles in the autism spectrum and other neurodevelopmental disorders. J. Autism Dev. Disord. 46, 2940–2955.

Mikadze, Y.V., 2014. The principles of plasticity in Lurian neuropsychology. Psychol. Neurosci. 7, 435–441.

Rangmar, J., Sandberg, A.D., Aronson, M., Fahlke, C., 2015. Cognitive and executive functions, social cognition and sense of coherence in adults with fetal alcohol syndrome. Nord. J. Psychiatry 69 (6), 1754–1760.

Rasmussen, C., Tamana, S., Baugh, L., Andrew, G., Tough, S., Zwaigenbaum, L., 2013. Neuropsychological impairments on the NEPSY-II among children with FASD. Child Neuropsychol. 19 (4), 337–349.

Roid, G.H. (2003). Stanford–Binet Intelligence Scales, 5th Edition (SB5). Riverside Publishing.

Roselló, R., Martínez-Raga, J., Mira, A., Girela, B., Cortese, S., 2021. Developmental outcomes in adolescence of children with autism spectrum disorder without intellectual disability: a systematic review of prospective studies. Neurosci. Biobehav. Rev. 126, 590–603. doi:10.1016/j.neubiorev.2021.04.010.

Rumball, F., Brook, L., Happé, F., Karl, A., 2021. Heightened risk of posttraumatic stress disorder in adults with autism spectrum disorder: the role of cumulative trauma and memory deficits. Res. Dev. Disabil. 110, 103848.WM difficulties may play a role in the susceptibility of youth with ASD to more severe reactions to trauma.

Sameroff, A., Chandler, M., 1975. Reproductive risk and the continuum of caretaking causality. In: Horowitz, F. (Ed.), Review of Child Development Research,Vol. 4. University of Chicago Press, Chicago, IL, pp. 187–243.

Sanders, A.F., Hobbs, D.A., Stephenson, D.D., Laird, R.D., Beaton, E.A., 2017. Working memory impairments in chromosome 22q11. 2 deletion syndrome: the roles of anxiety and stress physiology. J. Autism Dev. Disord. 47, 992–1005.

Schuiringa, H., van Nieuwenhuijzen, M., Orobio de Castro, B., Matthys, W., 2017. Executive functions and processing speed in children with mild to borderline intellectual disabilities and externalizing behavior problems. Child Neuropsychol. 23 (4), 442–462.

van Nieuwenhuijzen, M., Orobio de Castro, B., Matthys, W., 2017. Executive functions and processing speed in children with mild to borderline intellectual disabilities and externalizing behavior problems. Child Neuropsychol. 23 (4), 442–462.

Wechsler, D., 2014. Wechsler Intelligence Scale for Children–Fifth Edition (WISC–V), NCS Pearson.

Wechsler, D. (2017). Wechsler Adult Intelligence Scale–Fifth Edition (WAIS–V), Pearson, Bloomington, MN.

Zelazo, P.D., Frye, D., Rapus, T., 1996. An age-related dissociation between knowing rules and using them. Cogn. Dev. 11 (1), 37–63.

CHAPTER 3

Medical conditions and aggressive behavior

I. Leslie Rubin[a,b], Angela Hassiotis[c], and Lauren Charlot[d]
[a]Department of Pediatrics, Morehouse School of Medicine, Atlanta, GA, United States
[b]Department of Pediatrics, Emory University, Atlanta, GA, United States
[c]University College London, London, United Kingdom
[d]UMass Chan Medical School, Worcester, MA, United States

As discussed in previous chapters, people with autism spectrum disorder (ASD) and other intellectual and developmental disabilities (IDD) have higher rates of aggressive behavior compared with people of a similar age who do not have these conditions. Bardoloi reported a median rate of aggressive behavior among people with ASD and other IDD of 20% (Bardoloi, 2023). Given past reports of high rates of medical issues suspected of causing aggressive behavior to the point of requiring inpatient psychiatric care, it is important to carefully assess people with ASD/IDD for medical conditions when they present with aggressive behavior in a mental health crisis (Charlot et al., 2011). However, the assessment for medical conditions that may be causing the aggressive behavior among people with ASD/IDD does not routinely occur (Allely et al., 2013).

> *"Disruptive behaviours displayed by patients seem to induce doctors to make diagnostic errors."*
>
> ***Schmidt et al., (2017)***

Physical distress associated with some medical problems, particularly those involving pain or discomfort, can cause anyone to become irritable. Individuals with ASD/IDD who experience pain or discomfort are at an increased risk of engaging in aggressive behavior, in part because of their limited ability to verbally express their feelings (Allely et al., 2013). Furthermore, with a limited ability to tolerate and modulate their experiences and feelings, there is a greater tendency for individuals with ASD/IDD to exhibit dramatic fright, flight, and fight responses

Aggressive Behavior of People with Autism and other Intellectual and Developmental Disabilities
https://doi.org/10.1016/B978-0-443-26552-5.00007-2

to even mild disturbances or disruptions in their lives (Corbett et al., 2016; Bradley and Korossy, 2022). Therefore medical conditions that cause pain or discomfort can aggravate existing challenging behavior patterns, such as aggression toward self, others, and/or property. A variety of medical conditions have been identified as causing challenging and aggressive behaviors among people with ASD/IDD, including ear infections, premenstrual pain, sleep disturbances, allergies, dental pain, seizures, and gastrointestinal distress, including gastroesophageal reflux disease (GERD) and constipation (Charlot et al., 2011; Edelson, 2022; Gunsett et al., 1989; Kildahl et al., 2024; Kennedy and Meyer, 1996; O'Reilly, 1997; Ooi et al., 2023; Kennedy et al., 2007; El-Tallawy et al., 2023; Carr and Owen-DeSchryver, 2007). In behavioral terms, pain or distress from a physical condition acts as a "setting event" or "establishing operation" (events or states that increase the probability of occurrence of problem behaviors such as aggressive behavior) (Carr and Owen-DeSchryver, 2007). In a treatment plan or behavior plan, clinicians should pay attention to sources of distress that are medical in nature. As noted in Chapter 9, data should be collected regarding observable indications of medical conditions that may cause discomfort or pain and require further medical assessment or intervention before they become serious enough to provoke a reaction that may manifest in an act of aggression. The aggressive reaction may elicit a response from health care providers that results in the administration of medications to control the behavior, without addressing the root cause, without sustained benefit, and with potential long-term adverse consequences (Bardoloi, 2023; Charlot et al., 2022).

Table 3.1 lists possible medical issues known to be associated with behavioral health concerns or their treatments. Details of observable signs of common concerns are offered; for example, in the case of GERD, a person may be burping a lot, becoming agitated after meals, regurgitating food, or complaining of a "stomachache." Possible drug-induced movement disorders are also listed in terms of what an observer may report. Any signs of concern can be tracked on a data form (as shown in Chapter 9). These conditions and the discomfort, distress, or pain they cause increase the risk for aggressive behavior, particularly for people who are unable to verbally report internal states accurately. Thus aggressive behavior can result from a caregiver's and/or clinician's failure to fully assess the potential for contributing medical conditions.

Joey, an 18-year-old male with ASD, was referred for assessment and was very aggressive, especially toward himself. He had lost his school placement and was on multiple psychotropic medications to the point of being slowed and sedated much of the time. It was discovered, during the process of taking his history, that his treatment team believed he had been making himself vomit to "get attention." It was recommended to have a GI workup, and it was discovered that Joey had severe esophageal erosion from acid reflux, which was probably the cause of his initial decompensation.

Table 3.1 Common medical conditions likely to cause discomfort, pain, or distress and aggressive behavior.***

Condition	What may be observed	Diagnosis and treatment
Constipation	May start with c/os diarrhea, loose stool that is leaking around a high bowel impaction, c/os of stomach aches, obsessing about the bathroom, staying in or going into the bathroom a lot, hard stools, and days without having a large bowel movement despite normal food intake	KUB with comment on stool content or careful charting Treatment varies; consider issues with cramping Reduce offending agents that are not clearly effective for targeted aggressive and other behaviors
Acid Reflux	Burping a lot, c/os of sore throat, upset stomach, holding chest area, vomiting, agitated behavior around meals—some people actually try to eat more to quell the acid sensation	Upper GI, but if difficult to do, might consider an empirical trial of a PPI, elevate the head of the bed, and stay upright after eating
UTIs and Urinary retention	May c/o of difficulty or discomfort when urinating, agitation related to the bathroom, or taking medications known to cause retention should raise suspicion	U/A, but might also need a test for retention
Orthostasis	"Sit-down strikes," flopping to the floor, look off balance when standing up from sitting or lying down	May need to change from the use of the offending agent if medication-induced. Assess for other causes of syncopal episodes
Ear infections	May see self hits to the head or hands held over ears; may just see general agitation and aggression	Person may have impaired hearing; routinely check
Dental issues	May see SIB to the head or face, decreased eating, and may just observe general agitation and aggression	Desensitization for anxiety; use of sedation dentistry, if needed

Continued

Table 3.1 Common medical conditions likely to cause discomfort, pain, or distress and aggressive behavior—cont'd

Condition	What may be observed	Diagnosis and treatment
Sleep apnea	Sleepy during the day, even when in bed all night Waking at night and poor sleep at night, snoring	May miss this and worsen it with psychotropics; need a sleep study and may need CPAP, BiPAP, or to elevate the head of the bed
Obesity	Weight gain, and also medication-related increased appetite and food drive, is seen as a behavioral "food-seeking" problem	Usually, miss the increased food drive and cause aggression by restricting access, then add medications, etc.
Skin rashes	May cause enough discomfort to lower the threshold for aggressive acts	Especially women suffering from yeast infections need antifungal treatment
Ear wax	Can cause poor hearing and lead to problems interpreted as noncompliance	Patient may be seen as noncompliant or even psychotic when confabulating
Fatigue and sedation	May cause frustration, irritability	Often causes nonmedical interventions to be less effective, with more aggression leading to yet more sedating medications
Motor slowing	May cause frustration, irritability, and is sometimes seen as refusing to respond, resisting directions, and task demands	May be accepted as it reduces the intensity of aggression, but it needs adjustments and multimodality treatment for the aggression
Ataxia, gait disturbances	May see agencies add staff without discovering the source; need to check if medications or neurological illness are causing this	May be more commonly a medication issue among patients treated with multiple psychotropics, but it is necessary to assess for neurological causes—rare, but consider, i.e., hydrocephalus

Table 3.1 Common medical conditions likely to cause discomfort, pain, or distress and aggressive behavior—cont'd

Condition	What may be observed	Diagnosis and treatment
Muscle stiffness neck, trunk, limbs	May see a person rubbing their neck, leaning to one side, with gait changes, minimal arm swing, and general agitation, as well as upset and aggression at times	Often, a medication needs adjustment and can cause EPS or med-induced Parkinsonism; this can be a driver of agitation and aggression
Swallow problem dysphagia	Coughing, choking, and aspiration pneumonias, caused in some cases by antipsychotic medications, may be treated by changes in diet, i.e., pureed foods or thickened liquids, which can cause frustration and aggression when the patient is unable to eat their preferred foods	May help to reduce or eliminate the offending agent, treat while tapering slowly, and progressively reintroduce foods from which the patient was restricted with improvement
Oculogyric crises	Caregivers ask why the person always stares at the ceiling, may question partial seizures, which cause significant discomfort and can be a driver of anxiety and aggression	Eliminate the offending agent, and if needed, switch to a non-antipsychotic intervention if the drug was being used to tamp down aggression
Tardive dyskinesia	Some patients report feeling embarrassed about tongue protrusions, which can be frustrating	Often, other movement problems are also present but may be subtle; it is important to address the offending agent when it is not clearly effective and needed
Akathisia	May be seen as a reason to add ADHD medications	Address the offending agent when it is not clearly effective and needed
Metabolic encephalo pathy	Can be seen as psychotic when reversible medication-related confusion, waxing and waning, is perceived as intentional behavior	May take longer than expected for mental status problems to resolve

c/os, complaints of; *U/A*, urinary analysis
***Examples and suggestions above are based on clinical assessments and interventions by the authors.

Despite the documented link between health concerns and problem behaviors, medical causes for agitated behavior are frequently missed in people with ASD/IDD. A major reason is the fact that many individuals with ASD/IDD are not able to verbally report or describe their own health issues (Walton et al., 2022; Buie et al., 2010). Despite limited verbal skills, it remains critical to ask the individual directly, attempt to elicit self-reports, and listen carefully and attentively to what the person may be able to tell us. Unfortunately, the reality is that the patients themselves may not ever be asked how or what they are feeling, or, when they do report discomfort, they are characterized as "attention-seeking." Another example is when people with ASD/IDD refuse to take prescribed medications; they may be characterized as *noncompliant*. In our experience on the inpatient unit at the University of Massachusetts Medical School (UMASS), medication refusals were sometimes due to the experience of side effects. In essence, therefore, the refusals to take the medication were effectively a "statement" by the patients of their experience related to taking the medication, not a case of being uncooperative.

Patient Mr. Y, for example, would not take his antipsychotic medications. He was seen as oppositional and defiant. After careful study, he was found to be experiencing pain in his neck from a dystonia that was caused by his medication. He showed improvement, including displaying less aggression, when the pain was relieved following medication adjustments.

In many instances, the history of the present illness is reported by caregivers or family members rather than by the patient directly (Abend and Silka, 1999; Cooper et al., 2004). In the absence of a direct personal report by the individual with ASD/IDD, careful and thoughtful observations of the individual's behavior can provide valuable clinical information about the person's feelings, discomfort, or pain.

Despite the fact that individuals with ASD/IDD have higher rates of medical problems than the general population, they often lack access to appropriate and effective health care (Cooper et al., 2004; Beange et al., 1995). When efforts are made to examine individuals more carefully using health checks or screens, previously unrecognized conditions have been identified at significant rates (Robertson et al., 2014; Cooper et al., 2004; Lennox et al., 2007).

Although there have been a significant number of reports on health problems among individuals with ASD/IDD, only a few specifically focus on the health problems that affect individuals receiving treatment for psychiatric or behavioral conditions (Kwok and Cheung, 2007). Davidson et al. (2003) studied the association between health status and behavior disorders in four age groups in a large cross-sectional investigation of adults with ASD/IDD and found that health problems were highly correlated with behavioral and emotional disturbances, including neurological disorders, cardiopulmonary disorders, asthma, visual and hearing impairments, gastrointestinal disorders, and neoplastic diseases. These authors argued that "occult" unidentified medical conditions may present as "treatment-resistant psychiatric illness" or behavioral disturbances, especially among individuals who have difficulty or are unable to describe their symptoms.

Kennedy et al. (2007) studied children with severe intellectual disabilities (ID) who had healthcare needs. Children with challenging behavior had twice as many healthcare needs when compared with peers who did not have challenging behavior. The authors suggested a clear need for multidisciplinary care coordination. Woo et al. (2003) found that 34% of older individuals from the general population admitted for psychiatric inpatient care had unrecognized medical disorders and emphasized the fact that medical disorders "may cause psychiatric symptoms." Unrecognized medical disorders identified among 79 consecutive admissions included constipation, urinary infections, and hypothyroidism. They concluded that elderly patients would likely suffer less severe mental health problems, or it might be possible to "even eliminate the mental disorder," by conducting more aggressive assessment and treatment of medical concerns.

Because of the potential for physical health problems to present as psychiatric disorders with challenging behavioral difficulties, Charlot et al. (2011) studied psychiatric inpatients with IDD to determine the prevalence of medical comorbidities. Given that longer hospital stays are disruptive to patients' lives, are associated with increased morbidity, and are more costly, the correlation between medical comorbidities and medication treatment with long lengths of hospital stay seemed to be worthy of further investigation. It was speculated that greater surveillance for medical problems may have prevented some hospital stays and shortened the length of others (Charlot et al., 2011). In an investigation of inpatients with IDD who received a comprehensive multidisciplinary assessment because their treatment was not improving their condition, we found high rates of medical problems and medication side effects (Charlot et al., 2022). Although the roles these medical conditions played in the ongoing mood and behavior challenges of the patients could not be directly correlated in this study, the commonly associated discomfort from many of these previously unrecognized conditions indicates a need for further investigation. For some conditions, usual tests or studies are not pursued, or reports fail to include the scope of possible clinically significant findings. For example, ordering an abdominal X-ray or kidneys, ureters, and bladder (KUB) to see if a person is constipated may be less common for young adult patients. Under typical circumstances, this may be overlooked by radiologists in a hospital setting, who would assume the request was sought to rule out more serious findings such as obstructions, volvulus, or other major conditions. It is helpful to write the order with a specific request for "comment on stool content," which would alert the radiologist to focus on, note, and report the bowel contents. Individuals with recurrent urinary tract infections (UTIs) may need assessment for possible urinary retention, which may not often be pursued in a young patient. In this situation, the use of medications such as benztropine, which may cause urinary retention, can lead to recurrent UTIs and urinary incontinence. In some cases, urinary incontinence may be presented by family members or direct care takers to assessing professionals as intentional or attention-seeking. At the same time, urinary incontinence is a potentially treatable condition.

Some clinical observations regarding elderly psychiatric patients may pertain to individuals with IDD as well, as both patient populations may be unable to articulate sources of distress clearly. Thus both groups are at risk of having undetected and undiagnosed medical problems. This may also be relevant for young neurotypical children, who may not be able to describe their symptoms and rely on the observation and interpretations of their parents or caregivers. However, in these situations, clinicians generally know they need to explore sources of problems that are not readily obvious on examination. In the case of adults, many conditions are diagnosed by taking a good history and asking questions about symptoms, including the history of onset and details of the complaints, which provide the clinician with key clues to the likely diagnosis and management.

For example, one of the inpatients developed a sudden onset of confusion, which the providers felt reflected a sign of schizophrenia in a 32-year-old man with mild ID (Charlot et al., 2011). *In his case, elevated epoxide levels from his anti-epilepsy drug (AED) turned out to be the culprit, causing a metabolic encephalopathy that resolved when switching the AED.*

In these cases, a diligent search for alternative possible causes of the clinical picture is critical, especially when the symptoms have an unusual presentation, such as a later age of onset or a more acute onset than is typical for the psychiatric symptomatology.

A strong association has been found between pain, anxiety, and aggression among patients with dementia (Sampson et al., 2015). The authors advised addressing pain as a way to reduce aggression and other symptoms of concern in this group. This may be similar to patients with ASD/IDD, who may not be able to accurately report pain or physical discomfort and, as a result, may become anxious or aggressive. "Their limited ability in describing and complaining about pain symptoms may lead to underdiagnosis of painful conditions" (Sampson et al., 2015).

It is also concerning that aggression displayed in response to a medical condition may be misunderstood (Allely et al., 2013; Buie et al., 2010; Liu et al., 2020). Children with ASD and aggression have been found to have significantly greater difficulties with gastrointestinal disorders than those without aggression. A systematic review "highlights the strong possibility that not all children with ASD express their physical discomfort in the same way as a neurotypical child would (i.e., cry, moan, seek comfort, etc.) which may lead caregivers and the medical profession to interpret this as pain insensitivity or incorrectly lead them to believe that the child is in 'no pain'" (Allely et al., 2013; Buie et al., 2010; Liu et al., 2020).

Epilepsy and seizure disorders are significantly more common in children and adults with IDD than in the general population, with prevalence estimates around 20%–25%, particularly in those with more severe disability (Robertson et al., 2015). Because both epilepsy and challenging behavior cluster in people with ASD/IDD, it has long been suspected that epilepsy may be specifically associated with aggression and other forms of challenging behavior; however, empirical findings are mixed. A secondary analysis of a large trial cohort of adults with IDD and established challenging behavior (n = 246) found no significant association between the presence of epilepsy and overall levels of challenging behavior after adjustment for confounders, suggesting that epilepsy per se is not a primary driver of these behaviors (Blickwedel et al., 2019). In contrast, other studies have reported modest but statistically significant elevations in aggression and self-injury among adults with IDD who have comorbid epilepsy compared with those without epilepsy (van Ool et al., 2016) and very high rates of aggressive/destructive, stereotyped, and self-injurious behavior in specialist epilepsy–IDD services, where 60%–63% of adults showed such behaviors and aggression was often rated as the most clinically problematic (Deb et al., 2020). The variability in the reviews highlights the reality that any association between epilepsy and aggression is likely to be indirect and multifactorial: shared underlying brain pathology, early-onset neurodevelopmental impairment, communication difficulties, autism, psychiatric comorbidity, and environmental stressors, all of which increase the risk of both epilepsy and challenging behavior (Kerr, 2016; van Ool et al., 2016). In addition, certain antiseizure medications (for example, commonly levetiracetam and, less commonly, others) can precipitate or exacerbate irritability and aggression in a vulnerable minority of individuals, adding a treatment-related pathway to behavioral disturbance (Brodie et al., 2016). Overall, the available evidence suggests that while epilepsy is an important marker of neurodevelopmental complexity in IDD and may be associated with higher rates of aggression in some clinical contexts, especially post-ictal states, aggressive challenging behavior usually reflects the cumulative impact of multiple biological, psychological, and environmental factors rather than epilepsy alone.

Because the drivers of irritability and aggression in individuals with ASD/IDD can be multifactorial, they must be addressed in a systematic, stepwise manner or in parallel to identify treatable contributors and institute appropriate management. A study of 58 adolescents with ASD who were admitted to a neurobehavioral unit for severe challenging behaviors suggested that 28% had medical concerns that may have contributed to their decompensation. Such findings support the need for a comprehensive medical assessment when patients with ASD/IDD present with irritability and aggression (Ooi et al., 2023).

There is a significantly increased risk of specific medical problems related to psychopharmacologic treatment, including obesity, hypertension, abnormalities in glucose regulation, and an array of associated health complications such as diabetes, hormonal dysregulation, and cardiovascular problems (de Kuijper et al., 2013; Sheehan et al., 2015; Smith et al., 2022). The occurrence of movement disorders has been recognized as a problem affecting people with IDD more so than others without IDD who are treated with psychotropic medications (Sheehan and Hassiotis, 2017). In a study focused on differential diagnoses of psychiatric disorders among patients with significant behavioral challenges, Charlot et al. (2020) found polypharmacy and medical comorbidity to be a common association. In their study, they found that GERD and constipation were frequently diagnosed, while many individuals experienced side effects frequently caused by psychotropic medications, including weight gain, sedation, swallowing problems, and movement problems such as akathisia, dystonia, and dyskinesias.

References

Abend, S.L., Silka,V.R., 1999. Medical care of patients with developmental disabilities. Ment. Health Asp. Dev. Disabil. 2, 68–70.

Allely, C.S., Gillberg, C., Minnis, H., 2013. Pain sensitivity and observer perception of pain in individuals with autistic spectrum disorder. Sci. World J. 2356-6140. 2013 (1). doi:10.1155/2013/916178.

Bardoloi, P.S., 2023. Management of aggressive behaviors in neurodevelopmental disorders-role of α 2 agonists. Acta Sci. Neurol. 2582-1121. 6 (2), 12–16.

Beange, H., McElduff, A., Baker, W., 1995. Medical disorders of adults with mental retardation: a population study. Am. J. Ment. Retard. 99 (6), 595–604. doi:10.1352/0895-8017(1995)099<0595:MDOAWR>2.0.CO;2.

Blickwedel, J., Vickerstaff, V., Walker, M., Hassiotis, A., 2019. Challenging behaviour, epilepsy and intellectual disability: a secondary analysis of findings from a randomised controlled trial. J. Intellect. Dev. Disabil. 1366-8250. 44 (4), 457–463. doi:10.3109/13668250.2019.1587594.

Bradley, E., Korossy, M., 2022. Are difficult behaviours described in intellectual and developmental disabilities and autism actually adaptive responses to feeling unsafe? J. Dev. Disabil. 27 (1), 1–10.

Brodie, M.J., Besag, F., Ettinger, A.B., Mula, M., Gobbi, G., 2016. Epilepsy, antiepileptic drugs, and aggression: an evidence-based review. J. Neurol. Neurosurg. Psychiatry 87 (9), 961–969. doi:10.1136/jnnp-2015-311371 . ScienceDirect.

Buie, T., Campbell, D.B., Fuchs, G.J., Furuta, G.T., Levy, J., Van De Water, J., Whitaker, A.H., Atkins, D., Bauman, M.L., Beaudet, A.L., Carr, E.G., Gershon, M.D., Hyman, S.L., Jirapinyo, P., Jyonouchi, H., Kooros, K., Kushak, R., Levitt, P., Levy, S.E., Lewis, J.D., Murray, K.F., Natowicz, M.R., Sabra, A., Wershil, B.K., Weston, S.C., Zeltzer, L., Winter, H., 2010. Evaluation, diagnosis, and treatment of gastrointestinal disorders in individuals with ASDs: a consensus report. Pediatrics. 1098-4275. 125 (1), S1–S18S1. http://pediatrics.aappublications.org/cgi/reprint/125/Supplement_1/S1. doi:10.1542/peds.2009-1878C.

Carr, E.G., Owen-DeSchryver, J.S., 2007. Physical illness, pain, and problem behavior in minimally verbal people with developmental disabilities. J. Autism Dev. Disord. 1573-3432. 37 (3), 413–424. doi:10.1007/s10803-006-0176-0.

Charlot, L., Abend, S., Ravin, P., Mastis, K., Hunt, A., Deutsch, C., 2011. Non-psychiatric health problems among psychiatric inpatients with intellectual disabilities. J. Intellect. Disabil. Res. 0964-2633. 55 (2), 199–209. doi:10.1111/j.1365-2788.2010.01294.x.

Charlot, L.R., Doerfler, L.A., McLaren, J.L., 2020. Psychotropic medications use and side effects of individuals with intellectual and developmental disabilities. J. Intellect. Disabil. Res. 0964-2633. 64 (11), 852–863. doi:10.1111/jir.12777.

Charlot, L.R., Hodge, S.M., Holland, A.L., Frazier, J.A., 2022. Psychiatric diagnostic dilemmas among people with intellectual and developmental disabilities. J. Intellect. Disabil. Res. 0964-2633. 66 (10), 805–816. doi:10.1111/jir.12972.

Cooper, S.-A., Melville, C., Morrison, J., 2004. People with intellectual disabilities: their health needs differ and need to be recognised and met. BMJ 329 (7463), 414–415. doi:10.1136/bmj.329.7463.414.

Corbett, B.A., Muscatello, R., Blain, S.D., 2016. Impact of sensory sensitivity on physiological stress response and social anxiety in children with autism spectrum disorder. J. Autism Dev. Disord. 46 (5), 1624–1637. doi:10.1007/s10803-015-2681-.

Davidson, P.W., Henderson, C.M., Janicki, M.P., Assefa, D.J., Carter, J., 2003. Health status, health conditions, and health behavior among older adults with intellectual disabilities. Aging Ment. Health 7 (6), 478–485. doi:10.1080/13607860310001594673.

de Kuijper, G., Hoekstra, P.J., Visser, F., Scholte, F.A., Penning, C., 2013. Effects of discontinuing long-term antipsychotic drug use in people with intellectual disability. J. Clin. Psychopharmacol. 33 (1), 8–14. doi:10.1097/JCP.0b013e31827d0ade.

Deb, S., Akrout Brizard, B., Limbu, B., 2020. Association between epilepsy and challenging behaviour in adults with intellectual disabilities: systematic review and meta-analysis. BJPsych Open. 2056-4724. 6 (5). Available from: https://www.cambridge.org/core/journals/bjpsych-open. doi:10.1192/bjo.2020.96.

Edelson, S.M., 2022. Understanding challenging behaviors in autism spectrum disorder: a multi-component, interdisciplinary model. J. Pers. Med. 2075-4426. 12 (7). Available from: http://www.mdpi.com/journal/jpm. doi:10.3390/jpm12071127.

El-Tallawy, S.N., Ahmed, R.S., Nagiub, M.S., 2023. Pain management in the most vulnerable intellectual disability: a review. Adis Saudi Arab. Pain Ther. 2193-651X. 12 (4), 939–961. Available from: https://www.springer.com/journal/40122. doi:10.1007/s40122-023-00526-w.

Gunsett, R.P., Mulick, J.A., Fernald, W.B., Martin, J.L., 1989. Brief report: indications for medical screening prior to behavioral programming for severely and profoundly mentally retarded clients. J. Autism Dev. Disord. 1573-3432. 19 (1), 167–172. doi:10.1007/BF02212728.

Kennedy, C.H., Juárez, A.P., Becker, A., Greenslade, K., Harvey, M.T., Sullivan, C., Tally, B., 2007. Children with severe developmental disabilities and behavioral disorders have increased special healthcare needs. Dev. Med. Child Neurol. 1469-8749. 49 (12), 926–930. doi:10.1111/j.1469-8749.2007.00926.x.

Kennedy, C.H., Meyer, K.A., 1996. Sleep deprivation, allergy symptoms, and negatively reinforced problem behavior. J. Appl. Behav. Anal. 0021-8855. 29 (1), 133–135. Available from: http://onlinelibrary.wiley.com/journal/10.1002/(ISSN)1938-3703. doi:10.1901/jaba.1996.29-133.

Kerr, M., 2016. Behavioral disorder in people with an intellectual disability and epilepsy. Curr. Opin. Psychiatry 29 (2), 97–101. doi:10.1097/YCO.0000000000000223.

Kildahl, A.N., Berggren, T., Rønneberg, A., Serina Yksnøy Molnes, C., Nordgarden, H., 2024. A call for trauma-informed dental care for individuals with intellectual disabilities. Spec. Care Dent. 0275-1879. 44 (4), 1126–1134. doi:10.1111/scd.12959.

Kwok, H., Cheung, P.W.H., 2007. Co-morbidity of psychiatric disorder and medical illness in people with intellectual disabilities. Curr. Opin. Psychiatry. 0951-7367. 20 (5), 443–449. doi:10.1097/YCO.0b013e3282ab9941.

Lennox, N., Bain, C., Rey-Conde, T., Purdie, D., Bush, R., Pandeya, N., 2007. Effects of a comprehensive health assessment programme for Australian adults with intellectual disability: a cluster randomized trial. Int. J. Epidemiol. 36 (1), 139–146. doi:10.1093/ije/dyl254.

Liu, J., Chen, L. L., Shen, S., Mao, J., Lopes, M., Liu, S., Kong, X., 2020. Challenges in the diagnosis and management of pain in individuals with autism spectrum disorder. J. Autism Dev. Disord. 50, 3909–3920. doi:10.1007/s10803-020-04420-3.

O'Reilly, M.F., 1997. Functional analysis of episodic self-injury correlated with recurrent otitis media. J. Appl. Behav. Anal. 0021-8855. 30 (1), 165–167. Available from: http://onlinelibrary.wiley.com/journal/10.1002/(ISSN)1938-3703. doi:10.1901/jaba.1997.30-165.

Ooi, A., Banno, B., McFee, K., Elbe, D., Friedlander, R., 2023. Evaluating and managing irritability and aggression in children and adolescents with autism spectrum disorder: an algorithm. Br. Columbia Med. J. 65(8).

Robertson, J., Hatton, C., Emerson, E., Baines, S., 2015. Prevalence of epilepsy among people with intellectual disabilities: a systematic review. Seizure. 1059-1311. 29, 46–62. doi:10.1016/j.seizure.2015.03.016.

Robertson, J., Roberts, H., Emerson, E., Turner, S., Greig, R., Hatton, C., 2014. The impact of health checks for people with intellectual disabilities: an updated systematic review of evidence. J. Appl. Res. Intellect. Disabil. 27 (1), 25–33. doi:10.1111/jar.12055.

Sampson, E.L., White, N., Lord, K., Leurent, B., Vickerstaff, V., Scott, S., Jones, L., 2015. Pain, agitation, and behavioural problems in people with dementia admitted to general hospital wards: a longitudinal cohort study. Pain. 1872-6623. 156 (4), 675–683. Available from: https://journals.lww.com/pain/Pages/currenttoc.aspx. doi:10.1097/j.pain.0000000000000095.

Schmidt, H.G., Van Gog, T., Schuit, S.C., Berge, K.V.D., Van Daele, P.L., Bueving, H., Van Der Zee, T., Van Den Broek, W.W., Van Saase, J.L.C.M., Mamede, S., 2017. Do patients' disruptive behaviours influence the accuracy of a doctor's diagnosis? A randomised experiment. BMJ Qual. Saf. 2044-5415. 26 (1), 19–23. Available from: http://qualitysafety.bmj.com/content/by/year. doi:10.1136/bmjqs-2015-004109.

Sheehan, R., Hassiotis, A., 2017. Reduction or discontinuation of antipsychotics for challenging behaviour in adults with intellectual disability: a systematic review. Lancet Psychiatry. 2215-0366. 4 (3), 238–256. doi:10.1016/s2215-0366(16)30191-2.

Sheehan, R., Hassiotis, A., Walters, K., Osborn, D., Strydom, A., Horsfall, L., 2015. Mental illness, challenging behaviour, and psychotropic drug prescribing in people with intellectual disability: UK population based cohort study. BMJ. 1756-1833. 351, h4326. doi:10.1136/bmj.h4326.

Smith, E., Willis, V., Adler, S., Brennan, A., McMahon, R., Bertelli, M.O., 2022. Antipsychotic-associated weight gain in adults with intellectual disabilities: a systematic review. Acta Psychiatr. Scand. 146 (4), 302–319. doi:10.1111/acps.13488.

van Ool, J.S., Snoeijen-Schouwenaars, F.M., Schelhaas, H.J., Tan, I.Y., Aldenkamp, A.P., Hendriksen, J.G.M., 2016. A systematic review of neuropsychiatric comorbidities in patients with both epilepsy and intellectual disability. Epilepsy Behav. 1525-5050. 60, 130–137. doi:10.1016/j.yebeh.2016.04.018.

Walton, K., Krahn, G.L., Buck, A., Andridge, R., Lecavalier, L., Hollway, J.A., Davies, D.K., Arnold, L.E., Havercamp, S.M., 2022. Putting "ME" into measurement: adapting self-report health measures for use with individuals with intellectual disability. Res. Dev. Disabil. 1873-3379. 128. Available from: www.elsevier.com/locate/redevdis. doi:10.1016/j.ridd.2022.104298.

Woo, B.K.P., Daly, J.W., Allen, E.C., Jeste, D.V., Sewell, D.D., 2003. Unrecognized medical disorders in older psychiatric inpatients in a senior behavioral health unit in a University Hospital. J. Geriatr. Psychiatry Neurol. 16 (2), 121–125. doi:10.1177/0891988703016002011.

CHAPTER 4

Genotypes, phenotypes, and aggressive behavior

I. Leslie Rubin[a,b] and Lauren Charlot[c]
[a]Department of Pediatrics, Morehouse School of Medicine, Atlanta, GA, United States
[b]Department of Pediatrics, Emory University, Atlanta, GA, United States
[c]UMass Chan Medical School, Worcester, MA, United States

Introduction

When assessing possible causes of neurodevelopmental disorders (NDD), it is helpful to consider what it means when we say that an individual displays characteristics of a given syndrome. A syndrome refers to a collection of physical, physiological, and functional characteristics that occur together in a recognizable pattern. Syndromes may have known or unknown causes and vary in expression from person to person.

NDD can be caused by genetic factors and/or environmental factors, which commonly operate prenatally, "in utero," during fetal development, especially in the context of brain development (Love et al., 2024; Millan, 2013; Doi et al., 2022). Genetic anomalies likely causing intellectual and developmental disabilities (IDD) may account for as much as 82% of cases (Karam et al., 2015; Rodan et al., 2025). Each genetic syndrome has specific features depending on the identified genetic anomaly. It is estimated that more than 1%–3% of the population has IDD, with growing numbers reported for autism spectrum disorder (ASD).

A brief history of the genetics of IDD

The study of genetics, initiated by the work of Gregor Mendel in the mid-19th century, has been followed by significant scientific and technical advancements, notably in the late 20th century. Since the mapping of the human genome, further dramatic advances over the past 30 years have led to the identification of previously unknown genetic causes of IDD. As a result, the number of identified genetic causes of ASD and other IDD has grown exponentially (Stenseth et al., 2022).

Aggressive Behavior of People with Autism and other Intellectual and Developmental Disabilities
https://doi.org/10.1016/B978-0-443-26552-5.00003-5

In the early 20th century, conditions like Down syndrome were recognized for their distinct physical features, characteristic facial profiles, neurological conditions such as hypotonia, and congenital anomalies, commonly heart disorders. Then, in 1959, came the discovery that individuals with Down syndrome had an extra chromosome 21—Trisomy 21. Fragile X syndrome was first identified in the late 1940s when researchers noticed a pattern of fragile sites on the X chromosome in individuals with intellectual disabilities (IDs). Over the intervening decades, advances in genetics and imaging technologies further clarified the connections between chromosomal anomalies and specific physical traits, as well as underlying IDs, thus enhancing diagnosis and understanding of various syndromes. The latter half of the 20th century saw a paradigm shift with the advent of molecular genetics. The mapping of the human genome began to reveal the genetic underpinnings of various NDDs. As technology advanced, the focus expanded from identifying chromosomal abnormalities to understanding the role of specific genes in neurodevelopmental and neurobehavioral patterns. The late 1990s and early 2000s ushered in the era of genome-wide association studies (GWAS) and chromosome microarray analysis (CMA), enabling the identification of copy number variations (CNVs) associated with NDDs. For example, deletions and duplications on chromosome 22q11 were found to be linked to DiGeorge syndrome, which presents with characteristic physical features, developmental delays, and behavioral challenges. Today, we have unprecedented sophisticated genetic technology that enables us to identify genetic links to multiple conditions, enhancing clinical assessment and management, genetic counseling and research strategies. These advancements are forging new and previously unimaginable opportunities for treatment and, more excitingly, for prevention.

Given the dramatic advances in genetic technology, along with ready access to sophisticated genetic testing, recognition of the role of genetics in the functional and behavioral characteristics of individuals with IDD is clear and becoming increasingly relevant. Genetics should, therefore, become a logical and indispensable part of the clinical assessment and management of individuals with IDD, especially those with medical complications and significant behavioral challenges.

Although we know genetic factors play a major role in ASD specifically, and IDD in general, there is significant clinical and genetic heterogeneity (see ASD, 2025). Genetic testing can lead to the identification of a specific syndrome that may have associated medical conditions and, at times, behavioral patterns or phenotypes. Understanding these associated features can help in the process of assessing and treating aggressive behavior. Even if there are reports of previous genetic testing, including those conducted only 10 years ago, it is advisable to obtain updated genetic testing because of the rapid and dramatic advances in genetic testing in recent years.

Current technologies and best practices in genetic testing

Whereas, in past decades, blood would be drawn and sent to the lab for chromosome analysis, over time, increasingly focused genetic tests were developed, such as fluorescence in situ hybridization (FISH), a technique used in genetic testing to help diagnose chromosomal abnormalities before or after birth, or chromosomal abnormalities in an embryo during preimplantation genetic testing as part of IVF (in vitro fertilization). Today, newer technology can harvest cells obtained from a buccal swab to perform a detailed analysis of the map of each chromosome. CMA is a technology that detects submicroscopic deletions and duplications that are often missed by conventional karyotyping and can identify pathogenic CNVs in 10%–30% of individuals with IDD (Rodan et al., 2025; Vissers et al., 2015; Wang et al., 2025). The diagnostic yield varies by phenotype: children with syndromic features (e.g., dysmorphism, congenital anomalies) and moderate-to-severe ID have the highest rates of pathogenic CNVs (up to 25%–30%), while those with isolated ASD or mild developmental delay have lower yields (3%–10%) (Sandoval-Talamantes et al., 2023; Chaves et al., 2024). It is generally accepted that CMA is a first-tier clinical diagnostic test for individuals with ASD or other expressions of IDD, especially when the etiology of the NDD is unknown, there are associated congenital anomalies, or the individual is not responding to standard treatment protocols.

Specific recurrent CNVs (e.g., 16p11.2, 22q11.2, 1q21.1, 15q11.2, 3q29) are associated with a spectrum of neurodevelopmental and psychiatric phenotypes, including ASD, ID, epilepsy, and schizophrenia, and can present with variable expressivity and penetrance—even within families (Birnbaum et al., 2022). The same CNV may manifest as ASD in one individual and as ID, epilepsy, or psychiatric disorders in another. Challenging behaviors, such as aggression, self-injury, and severe social impairment, are frequently observed in individuals with pathogenic CNVs, especially those affecting genes involved in synaptic function and neurodevelopment (Birnbaum et al., 2022).

Next-generation sequencing (NGS) refers to high-throughput DNA sequencing technologies that allow rapid, comprehensive analysis of genetic variation across the genome, and it is increasingly used to identify the genetic causes of NDDs, including chromosomal abnormalities and single-gene mutations. NGS technology simultaneously sequences millions of short DNA or RNA fragments to determine the order of nucleotides. This "massively parallel" approach has revolutionized genomics by significantly increasing speed, decreasing costs, and enabling high-throughput analysis of entire genomes, targeted gene regions, gene expression, and genetic variations. These advancements have led to widespread applications in research and clinical settings, such as disease diagnosis, personalized medicine, and cancer research (Khan et al., 2025).

NGS encompasses several approaches:

- *Exome sequencing* targets all protein-coding regions (exons) of the genome, where most disease-causing mutations are found. It is highly effective for diagnosing monogenic causes of NDDs, with a diagnostic yield of 28%–43% in ID/global developmental delay (ID/GDD) and ASD (Rodan et al., 2025; Alverez-Mora et al., 2022).
- *Genome sequencing* analyzes nearly the entire genome, including coding and noncoding regions, and can detect a broader range of variant types, such as single nucleotide variants, small insertions/deletions, and structural variants (e.g., large deletions, duplications, translocations). Genome sequencing offers a 10%–20% higher diagnostic yield than exome sequencing and can resolve complex or noncoding variants missed by other methods (Sanchis-Juan et al., 2023; Shin et al., 2023; van der Sanden et al., 2023).

Much genetic research on ASD is now focused on "multi-gene" risk factors. NGS approaches, in combination with new functional readout systems, are revealing more about the genetic errors in our biological makeup that affect organ system function, especially in the brain. Knowing more about the biological pathways underlying NDDs has improved the diagnostic process and offered new targets for therapy. The emerging studies are identifying many more genes that may be involved in or contribute significant risk to ASD/IDD. For example, when one identical twin has ASD, the likelihood that the other twin will also have this condition can be greater than 90%, while about 34% of fraternal twins of the same sex meet the criteria for ASD (Cox, n.d.). In some cases, single CNVs are believed to be the cause of ASD, while in other cases, some "common" genetic variants increase risk but are insufficient by themselves to determine an ASD phenotype. Rare genomic variants, of various sizes, inherited from a parent or occurring de novo, can also be associated with the ASD phenotype (La Monica et al., 2025, p. 228). See the discussion below regarding epigenetics.

Inheritance patterns

Genetic anomalies may be inherited from one or both parents, or may arise de novo, that is, without any corresponding genetic anomaly in either parent. Inherited genetic disorders may be autosomal dominant, in which one of the parental chromosomes contributes to the individual's genetic makeup and causes the particular condition, or autosomal recessive, in which genes on chromosomes from both parents contribute to the genetic disorder. X-linked genetic disorders are conditions caused by mutations in genes located on the X chromosome. They are inherited in a sex-linked pattern and are more likely to affect males who inherit the mutated gene from their mother. The anomalies that arise de novo are most commonly CNVs (Lopes-Marques et al., 2024), and possibly one-third of unknown cases of severe IDD may be explained by CNVs.

Defining genotype and phenotype

A genotype is an individual's complete set of genes, which serves as the blueprint for the individual's observable physical and functional expression of those genes. Understanding specific genotypes informs us about how genetic makeup impacts an individual's physical, physiological, and functional domains. An individual's phenotype is defined as a set of observable and measurable characteristics, such as physical appearance, biochemical properties, and behavior of that individual. It is important to note that the genotype is sensitive and responsive to environmental factors, which can shape the phenotype.

A behavioral phenotype is a characteristic and consistent pattern of cognitive, personality, or behavioral traits associated with a specific genetic disorder. These patterns, which can include social, communication, and motor profiles, are not static and can change, varying with age, intellectual ability, active intervention, and life experiences. Thus behavioral manifestations arise from the interaction of genetic makeup and environmental factors; this classic "nature-nurture" interaction often includes a wide variety of symptoms.

Complex models are required to explain behavioral outcomes, such as who develops psychiatric disorders and who will become dangerously aggressive (see detailed discussions in Chapters 5 and 9). It is now understood that these outcomes can be viewed as a product of gene-environment interactions or transactions. For example, transactions between the child and their environment can be considered as "evocative correlations." An evocative correlation is a relationship where a person's genetically influenced traits elicit specific reactions from their environment in a kind of feedback loop (Klahr et al., 2013). It is also well established that adverse childhood experiences (ACEs) can result in aggressive and violent behavior in adults (Burke et al., 2023). Individuals with ASD/IDD may have had ACEs, which can play a significant role in patterns of aggressive behavior. Over time, there are a variety of mediators and moderators, as well as risk and protective factors, that influence eventual behavior, personality, and phenotypic outcomes. Thus, when considering how genes impact outcomes, it is helpful to view genes as providing a blueprint for who we may become (our genotype), while life experiences influence final outcomes or phenotypes.

(See Tables 4.1–4.8, where several syndromes known to produce neurodevelopmental conditions are described, including ways in which commonly associated features may increase the risk of aggressive behavior.)

Epigenetics

Abnormal patterns in our genes can cause problems in brain development. Epigenetics is the study of how genes are expressed when changes occur without alterations in the DNA sequences, and it addresses how interactions between genes and environments impact developing brains. Epigenetics acts as an interface between genetic predisposition and environmental exposures, regulating gene expression without altering the DNA sequence. During brain development, precise epigenetic regulation is essential for neuronal differentiation, maturation, and circuit formation. Aberrant epigenetic markers can disrupt these processes, leading to altered neuronal identity, connectivity, and function, which underlie neurodevelopmental disabilities (Reichard and Zimmer-Bensch, 2021; Love et al., 2024; Millan, 2013). Epigenetic mechanisms, including DNA methylation, histone modification, and chromatin remodeling, play a critical role in neurodevelopment, and disruptions in these processes are increasingly linked to neurodevelopmental disabilities such as ASD, ID, ADHD, schizophrenia, and various genetic syndromes. In ASD, genome-wide changes in DNA methylation and histone acetylation have been identified, affecting genes involved in synaptic function, neuronal excitability, and immune responses (LaSalle, 2023; Tseng et al., 2022). Genetic syndromes such as Rett syndrome, fragile X syndrome, and Prader–Willi syndrome are characterized by mutations in genes that regulate epigenetic processes, leading to widespread transcriptional dysregulation (Rangasamy et al., 2013). The available evidence shows that genetic susceptibility to conditions like schizophrenia and ASD is associated with distinct DNA methylation patterns detectable at birth, suggesting early epigenetic involvement in disease risk (Schuurmans et al., 2024).

The structural impact of genetic anomalies, in some instances, results in traits that are further shaped by environmental influences. Epigenetically associated changes during prenatal or early postnatal life can affect neurodevelopmental processes and outcomes. For example, maternal stress, environmental toxins, medications (e.g., valproic acid), alcohol use, infections, or nutritional compromise can all provoke epigenetic changes that increase the risk of ASD/IDD (Balachandar et al., 2022; Herrera et al., 2024). It was noted that COVID-19 during pregnancy can increase the risk of ASD (Shook et al., 2025).

A variety of mechanisms play a role in epigenetic effects. There can be an impact on synapse formation and brain plasticity. DNA methylation, a process responsible for the activation and "silencing" of certain genes, for example, is one of these mechanisms. Histone modification alters how tightly DNA is wrapped, affecting gene accessibility or availability for transcription. Gene expression can vary in relation to histone modification (the degree or extent of this). Gene activity at synapses is also regulated by noncoding RNAs that affect gene expression but do not code for proteins.

Examples of disorders resulting from epigenetic regulation problems include Rett syndrome (MECP2 mutation) and Fragile X. Epigenetic dysregulation affects stress response systems. For example, the hypothalamic-pituitary axis (HPA) may be involved in this process, as are aspects of neural circuitry such as dopaminergic and serotonergic pathways. Prefrontal cortical circuits that subserve impulse control and the effects listed above are all sources of influence on the aggressive behavior of people with ASD/IDD. The influences of genetic anomalies associated with ASD/IDD that contribute to the risk for aggressive behavior operate through the neurodevelopmental features they subserve, such as when there are executive function challenges. These may cause transactions that are stressful when a person has difficulty shifting set and is pushed by caregivers. Risks for aggressive behavior include exposure to stressful environmental inputs, tendencies to easily become over or under-aroused, and having fewer tools for managing them. These challenges, commonly seen in ASD/IDD, include sensory sensitivities, a lack of social awareness, difficulties in information processing and planning, and associated tendencies to become emotionally dysregulated.

Research has demonstrated the potential need to consider variants in epigenetic/chromatin regulatory genes (EC), which control gene expression by altering chromatin structure without changing the underlying DNA sequence, as having some influence over the risk for medication side effects. Lenz and colleagues studied over 300 youth and contrasted those with EC variants, those with other rare genetic disorders, and those without a variant of this kind. The children who had genetic anomalies were more likely to have ID, had more physical health conditions, and were more likely to experience medication side effects, including drowsiness/sedation and weight gain (Lenz et al., 2025).

Genetic syndromes such as Rett syndrome, fragile X syndrome, and Prader–Willi syndrome are characterized by mutations in genes that regulate epigenetic processes, leading to widespread transcriptional dysregulation. Scientists who study genes focus on events occurring at a molecular level. For example, a mutation in the FMR1 gene on the X chromosome results in an abnormal expansion or repetition of CGG sequences in excess of 200. This excessive repetition "silences" the FMR1 gene, preventing it from producing the FMRP protein. This protein has been determined to be important in normal brain development. In the case of Fragile X syndrome, for example, there are efforts underway to determine how to target the causes of the silencing of the FMR1 gene, with a focus on histone processes and methylation effects (Dall'Aglio et al., 2018; Schuurmans et al., 2024). Epigenetic changes are often tissue- and developmental stage-specific, complicating biomarker discovery and therapeutic targeting, but advances in epigenomic profiling hold promise for early diagnosis and intervention (LaSalle, 2023).

Epigenetic factors represent the biological component that combines with early developmental influences responsible for variability in the developmental profiles of people with ASD/IDD in many cases. Different people with IDD/ASD experience varied degrees of difficulty regulating their own behavior, with challenges related to coping with stress, which

Table 4.1 Smith–Magenis syndrome (SMS).

Causes	Physical features	Cognitive and other features	Behavioral phenotype	Links to aggressive behavior
Deletion of the short arm of chromosome 17 (17p11.2) most often affects RAI1 gene, which produces a protein that regulates other genes, impacts embryonic development, and serves as a key regulator of the circadian clock	Distinct facial features include a broad forehead, a flattened midface, a short upturned nose, and a tent-shaped upper lip. Additional characteristics include hypotonia (low muscle tone), scoliosis, vision problems (e.g., strabismus), and middle ear abnormalities. Sleep disturbances, such as difficulty falling asleep, nighttime waking, and excessive daytime sleepiness, are linked to abnormal melatonin secretion patterns.	Mild-to-moderate ID challenges in sensory sensitivities.	Often, individuals may exhibit many ASD features, attention challenges, and hyper motor activity. Hyperphagia, behavioral concerns such as self-biting, head banging, skin picking, and inserting objects into body openings (polyembolokoilamania), as well as involuntary spasmodic "self-hugs" or upper body squeezes, may also be observed. Some of these features may present as compulsions.	Similar to many individuals with challenges associated with ASD, anxiety likely plays a key role in increasing the probability of aggressive behavior when stress occurs. Increases in repetitive behaviors may indicate an increased risk for aggressive behavior Over-reliance on medications that disrupt sleep architecture or increase appetite can add to the risk of aggressive behavior.

From Korteling, D., Musch, J.L., Zinkstok, J.R., Boot, E., 2024. Psychiatric and neurological manifestations in adults with Smith–Magenis syndrome: a scoping review. Am. J. Med. Genet. B Neuropsychiatr. Genet. 195 (2), e32956.

Table 4.2 Williams syndrome.

Causes	Physical features	Medical and other features	Behavioral phenotype	Links to aggressive behavior
Microdeletion on chromosome 7 (7q11.23): The deleted region includes 26–28 genes; the deletions involve CLIP2, GTF2I, GTF2IRD1	"Elfin" facial appearance Small, upturned nose—long ridges in the skin run from the nose to the upper lip Prominent lips with an open mouth	Cardiac conditions: supravalvular aortic stenosis (SVAS)	Elevated risk for anxiety, ADHD, and phobias: approximately 50% meet the criteria for an anxiety disorder	Attention problems and anxiety are common mental health challenges. Aggressive behavior, when it occurs, is likely due to anxiety. Anxiety is usually at the core of any kind of challenging behaviors. Check on what might be causing anxiety; be supportive and reassuring. Music is a great way to promote wellness for many people, but especially for these individuals.
LIMK1 > challenges with visual-spatial tasks, unique behavioral characteristics, and other cognitive difficulties	Short stature	Narrowing of the large blood vessel that carries blood out from the heart	Heightened attraction to faces and a significant motivation toward social interactions	In emerging research, anxiety was linked to impairments in both social and executive functioning.

Table 4.2 Williams syndrome—cont'd

Causes	Physical features	Medical and other features	Behavioral phenotype	Links to aggressive behavior
Loss of GTF2IRD1—distinctive facial features	Skin that covers the inner corner of the eye (eye folds or epicanthal folds)	Muscle tone problems joint problems-stiffness	Very friendly and even too trusting of strangers, "chatty."	People with Williams syndrome often have special musical talents and are very friendly. There is a band composed of individuals who have Williams syndrome, along with some peers who have other disabilities, from Berkshire Hills Music Academy. Check out some of their music videos: https://www.youtube.com/channel/UCDmVkOBKCyTLe880DnCRznQ
	Partially missing teeth, defective tooth enamel, or small, widely spaced teeth.	Farsighted	Lower scores on visuospatial measures compared to verbal measures of cognition	Note that speaking and conversing, as well as social skills, are significant strengths. To prevent challenges: Avoid places with loud noise and provide adequate supervision—people with Williams syndrome may trust strangers when they should not because they are so friendly.
	Sunken chest (Pectus Excavatum)	High calcium (hyper-calcemia) early on		

From Royston, R., Oliver, C., Howlin, P., Waite, J., 2021. Anxiety characteristics in individuals with Williams syndrome. J. Appl. Res. Intellect. Disabil. 34 (4), 1098–1107.

Table 4.3 Cornelia de Lange syndrome (Prevalence: 1 in 10,000 live births).

Causes	Physical features	Medical & other features	Behavioral phenotype	Links to aggressive behavior
Mutation in the gene NIPBL occurs in 50% of cases and usually arises "de novo"	Short stature Low-set ears	High rates of gastroesophageal reflux disease (GERD)	Frequently associated with ASD	Rapid decreases are followed after Nissen fundoplication or, i.e., PPI.
Other gene abnormalities are associated with less severe forms and milder phenotypes, arising from mosaic genetic patterns that are more challenging to diagnose	Small, widely spaced teeth, long smooth philtrum, and downturned corners of the mouth	Gastrointestinal findings also include volvulus, rumination, chronic constipation, submucous cleft palate, sinusitis, and nasal polyps, as well as congenital diaphragmatic hernia	Often have sensory sensitivities	Cannot explain the SIB simply as a "learned behavior." Comprehensive medical screening is needed. Increases in challenging behaviors are often caused by medical issues, especially GERD and constipation.
NIPBL gene is responsible for producing the protein delangin. In utero, delangin plays a role in the development of arms and legs, the bones of the skull and face, the spinal column, and the heart, "cohesion complex mutations"	Small upturned nose	Pre and postnatal growth retardation	Sleep disturbances	Aggressive behavior is usually ANXIETY driven, often sensory-related, but also influenced by difficulty with new things, new people, and challenges to social cognition, which drive the risk for aggressive behaviors.

Table 4.3 Cornelia de Lange syndrome (Prevalence: 1 in 10,000 live births)—cont'd

Causes	Physical features	Medical & other features	Behavioral phenotype	Links to aggressive behavior
Clinical heterogeneity variable presentation	Hirsutism Long eyelashes Thick eyebrows	Musculoskeletal abnormalities of the upper extremities, including small hands and feet, clinodactyly of the fifth finger, and a missing forearm with digits	Self-injurious behavior	Swinging, therapeutic horseback riding, and swimming activities help with sensory challenges. Incorporate deep pressure rather than light touch. Use visual aids or pictures to assist with communication, or leverage technology (e.g., iPads). Reduce anxiety by introducing new things gradually and at a slow pace. Implement a predictable, structured visual daily schedule.

From Dai, Y.L., Qin, Y.F., Chao, Y.Q., Hu, C.X., Xia, F.L., Zou, C.C., 2024. The genotype and phenotype correlation of Prader-Willi syndrome. Rare Dis. Orphan Drugs J. 3 (4), N–A.

Table 4.4 Prader–Willi syndrome (PWS) (Prevalence: 1 in 15,000).

Causes	Physical features	Medical and other features	Behavioral phenotype	Links to aggressive behavior
Anomaly on chromosome 15, usually involving deletions in the 15q11.2-q13 region	Obesity affects 40% of children and adolescents and 80%–90% of adults	Hormonal deficiencies and hypogonadism (underdeveloped genitals)	ASD features include many repetitive and obsessive behaviors; psychosis, delusions, and hallucinations often emerge during adolescence or early adulthood; between 11% and 33% of individuals with mUPD are affected to a greater extent	Limit-setting on driven food-seeking and self-harm can lead to aggressive behavior toward oneself and others; a need for balanced approach focused on promoting positive engagement
Deletion with only one maternal copy accounts for 65%–75%	Short stature	Bone density problems, hypotonia (low muscle tone)	Obsessive-compulsive disorder and mood disorders, including social anxiety	Often, they need carefully controlled diets and restrictions on access to food.
Maternal uniparental disomy (mUPD) involves two maternal copies and occurs in 20%–30% of cases	Almond-shaped eyes, small chin	Sleep apnea	The obsession with food is so strong that some people with PWS will even eat food out of the garbage or consume frozen foods	Be sensitive to the fact that a person with PWS cannot help their desire for more food (their brains do not signal that they are full, and they constantly feel hungry)

Table 4.4 Prader–Willi syndrome (PWS) (Prevalence: 1 in 15,000)—cont'd

Causes	Physical features	Medical and other features	Behavioral phenotype	Links to aggressive behavior
Most common paternal chromosome 15 genetic material (although present) is inactive	Hypotonia (reduced muscle tone)	Other consequences of obesity in adults (e.g., high blood pressure, diabetes)	May eat to a level of danger.	Psychoactive medications can make this even worse; missing sleep apnea can be a factor that leads to irritability and aggressive behavior
Above involves abnormalities in the oxytocin system; clinical trials are equivocal for treatment of core symptoms		Decreased vomiting	High pain threshold	Promote mental wellness and reduce anxiety
This work examines combined targets with multiple modalities for treatment		Seizures	Skin picking can be severe	Use tactics that help with OCD and ASD, i.e., predictable routines, frequent small healthy snacks, and exercise
			Difficulty with changes in routine	Positive behavior supports focus on strengths and interests to engage, have fun, and divert focus from food or other obsessive tendencies, such as hoarding
				Never use food as a reward

From Dai,Y.L., Qin,Y.F., Chao,Y.Q., Hu, C.X., Xia, F.L., Zou, C.C., 2024. The genotype and phenotype correlation of Prader–Willi syndrome. Rare Dis. Orphan Drugs J. 3 (4), N–A; Godler, D.E., Singh, D., Butler, M.G., 2025. Genetics of Prader–Willi and Angelman syndromes: 2024 update. Curr. Opin. Psychiatry 38 (2), 95–100.

Table 4.5 22q11.2 deletion syndrome (Prevalence: 1 in 3,000–6,000 newborns 22q11.2DS—the most frequent microdeletion syndrome in the human genome).

Causes	Physical features	Medical and other features	Behavioral phenotype	Links to aggressive behavior
In 90% of cases, the 22q11.2 deletion occurs de novo during gametogenesis as ⋙ nonallelic homologous recombination events, while 10% of cases follow an autosomal-dominant inheritance pattern.	Underdeveloped chin Low-set ears	Congenital heart defects, arrhythmias, and aortic root dilation, as well as lifelong risk factors for CHDs, including hypertension, obesity, and hyperlipidemia, are significant concerns. Problematic thyroid function, with 20% experiencing hypothyroidism, is also noteworthy.	Extreme variability in the type and severity of phenotypes, including neonatal seizures and hypotonia, are very early signs of the disorder. Anxiety, schizophrenia/psychosis, behavioral disorders, and Parkinson's disease are also associated symptoms.	When someone you support has 22 q11.2 deletion syndrome, always refer them to their medical doctors.
Loss of another gene, COMT, in the same region as chromosome 22.	Cognitive: Prevalence > 90% for motor/speech delays and learning disabilities. Not all individuals have intellectual disabilities.	Absence of the parathyroid gland: 50%–65% of patients may experience hypocalcemia (symptoms include paresthesia, muscle spasms, cramps, and seizures).	The high rate of ASD is 14%–25%.	Initial medical "red flags" include low energy, breathing problems, changes in bladder habits, or drinking more fluids than usual.

Table 4.5 22q11.2 deletion syndrome (Prevalence: 1 in 3,000–6,000 newborns 22q11.2DS—the most frequent microdeletion syndrome in the human genome)—cont'd

Causes	Physical features	Medical and other features	Behavioral phenotype	Links to aggressive behavior
	Challenges in executive function, working memory, visuospatial abilities, motor skills, and social cognition/ social skills	Immunodeficiency, most notably moderate T-cell deficiency, is defined as having partial DiGeorge syndrome (pDGS).	Children are at an increased risk for ADHD, anxiety, and depression.	Though psychosis is more common in adults, sometimes people are depressed or have bipolar disorder.
		Palatal abnormalities and velopharyngeal dysfunction affect almost 65% of patients, while 11% have a cleft palate	Adults risk for schizophrenia, psychotic disorders, and bipolar disorder	There may be issues with drug treatment (i.e., if thrombocytopenia is an issue or due to renal problems).
		Renal congenital anomalies (CAKUT), which affect 30% of the urinary tract, include conditions such as bilateral or unilateral renal agenesis, duplicated collecting systems, and hydronephrosis. Other associated anomalies include cryptorchidism, hypospadias, absent uterus, or inguinal hernia. Thrombocytopenia risk		

From Szczawińska-Popłonyk, A., 2023. Chromosome 22q11.2 deletion syndrome. Int. J. Mol. Sci. 24 (9), 8317.

Table 4.6 Angelman syndrome (Prevalence: approximately 1 in 12,000 to 1 in 20,000 live births).

Causes	Physical and cognitive features	Behavioral phenotype	Links to aggressive behavior	Strategies to promote wellness and prevent aggressive acts
Mutations in the UBE3A gene on chromosome 15 prevent the production of a protein called E3 ubiquitin-protein ligase.	Hair, skin, and eyes that are light in color (hypopigmentation), scoliosis is common, children appear thin with low or near-normal weight, and delayed development. Severe speech impairment, epilepsy (in most cases), microcephaly (in most cases), and ataxia (movement and balance problems).	A happy demeanor with prominent "smiling." Laughing seems out of context. General exuberance associated with hyper motor Reduced behavioral adaptive skills.	Primary risk for autistic features may add risk for aggressive behavior. Individuals may have significant challenges with functional communication. People often assume the person is well or okay due to a disconnect between affective expression and events. Frustration can arise if the need for assistance isn't understood. Bullying and rejection at school are also risks.	

From Godler, D.E., Singh, D., Butler, M.G., 2025. Genetics of Prader–Willi and Angelman syndromes: 2024 update. Curr. Opin. Psychiatry 38 (2), 95–100.

Table 4.7 Fragile X (Prevalence: approximately 1 in 5,000 males and 1 in 4,000 to 1 in 8,000 females).

Causes	Physical features	Medical and other features	Behavioral phenotype	Links to aggressive behavior
Mutation in FMR1 gene on the X chromosome results in an abnormal expansion of CGG repeats—exceeding 200—which "silences" FMR1 gene's production of the FMRP protein.	Macrocephaly: 50%–81%; Long face: 83%; prominent ears; prominent jaw (in adults): 80%	Strabismus 8%–30%	Attention-deficit hyperactivity disorder: Anxiety 58%–86%, Depression30%–60%	Frustration regarding communication is related to many ASD features—difficulty with change, new people, and new places, as well as sensory issues.
≫ Dysregulation-overexpression imbalances of neurotransmission, along with FMRP inactivation, interfere with dendritic spine formation via alterations of neural mRNAs, which may reduce neural plasticity.		Recurrent otitis 47%–75%	Autism spectrum disorder affects approximately 80% of boys and 40% of girls who exhibit symptoms, which can sometimes be mistaken for shyness.	Sleep problems are common and can lead to irritability, increasing the risk of aggressive behavior.
Most common inherited cause of ID and of X-linked IDD.	High-arched palate	Gastrointestinal complaints 31%	Gaze aversion	Anxiety plays a key role in the risk of aggressive behaviors—anxiety reduction prevention reduces aggressive behavior.

Continued

Table 4.7 Fragile X (Prevalence: approximately 1 in 5,000 males and 1 in 4,000 to 1 in 8,000 females) —cont'd

Causes	Physical features	Medical and other features	Behavioral phenotype	Links to aggressive behavior
Individuals with a premutation, especially males, are at risk for fragile X-associated tremor/ataxia syndrome (FXTAS), while females with a premutation are at risk for fragile X-associated primary ovarian insufficiency (FXPOI) before the age of 40.	Flat feet	Obesity 30%–61%	Sadness or depression	Being prone to becoming easily overaroused increases the risk of aggressive behavior.
New sequencing has identified different mutations that cause FXS, including point mutations, deletions, and duplications.	Soft skin	Seizures 15%–20%	Tendency for sensory over arousal	Multiple modalities help reduce the effects of neurodevelopmental challenges, i.e., SLPs and OT.
No ID is present in some individuals with the mutation, while others exhibit mild to moderate ID. Approximately 85% of boys and 25%–30% of girls are affected. All boys and two-thirds of girls experience language challenges.	Other connective tissue abnormalities		Inattention, hyperactivity	
CGG triplets in the 5′ untranslated region of *FMR1* are hypermethylated.			Attention Deficit Disorder	

From Protic, D.D., Aishworiya, R., Salcedo-Arellano, M.J., Tang, S.J., Milisavljevic, J., Mitrovic, F., Budimirovic, D.B., 2022. Fragile X syndrome: from molecular aspect to clinical treatment. Int. J. Mol. Sci. 23 (4), 1935; Rajaratnam, A., Shergill, J., Salcedo-Arellano, M., Saldarriaga, W., Duan, X., Hagerman, R., 2017. Fragile X syndrome and fragile X-associated disorders. F1000Research 6, 2112..

Table 4.8 Down syndrome (Prevalence: global average 1 in 700–1000 live births).

Causes	Physical features	Medical and other features	Behavioral phenotype	Links to aggressive behavior
Trisomy 21 chromosomal anomaly	Short stature Hypotonia	Congenital heart defects (50%), thyroid disease (24%–50%), and Hashimoto thyroiditis (13%–39%)	Friendly, very social, often performs better in practical tasks compared to verbal IQ; mild to moderate ilityID	Usual links to aggressive behavior: frustration regarding communication and confusion.
96% are sporadic due to nonfamilial trisomy 21, in which there are 47 chromosomes with the presence of a free extra chromosome 21.	Small, brachycephalic head; small mouth; small ears; flat nasal bridge; excessive skin at the nape of the neck	Hearing loss (75%)Otitis media (50%–70%)Obstructive sleep apnea (50%–79%)	At times described as inflexible (may "dig their heels in" and be unwilling to change their opinions)	Provide reminders and visual cues to support learning and reduce frustrations
3% to 4% extra chromosomal material results from unbalanced translocation at chromosomes 14 or 21	Epicanthal folds, upward-slanting palpebral fissures, Brushfield spots	Eye problems (60%–80%), including cataracts, nasolacrimal duct obstruction (3%–36%), strabismus, and severe refractive errors (36%–80%)	Over a lifetime, 50% of people will experience a psychiatric disorder	Always give choices, even if it's just between two options, as this helps them feel a sense of control and see persistence as a strength when possible

Continued

Table 4.8 Down syndrome (Prevalence: global average 1 in 700–1000 live births)—cont'd

Causes	Physical features	Medical and other features	Behavioral phenotype	Links to aggressive behavior
	Wide spacing between the first and second toes, often accompanied by a deep plantar groove	Gastrointestinal atresia (12%), 6, 7, 8, and, less commonly, transient abnormal myelopoiesis (4%–10%)	Most common disorders: = Depression, Anxiety, OCD	See "oppositional" behavior as a form of communication expressing frustration, and try to support the person in better expressing their needs
	Single transverse palmar crease, short fifth finger with clinodactyly	Amyloid plaques and tau neurofibrillary tangles are virtually universal by age 40 years, and the lifetime risk of developing dementia is over 90% (Fortea et al., 2021); the mean age of onset for Alzheimer's diseaseis approximately 51–55 years.	Some ADHD and externalizing behavior challenges	Be sure the individual knows you genuinely care about their opinion
		Hip dislocation, hip abnormalities (2%–8%), and symptomatic atlantoaxial instability (1%–2%)	Fewer psychiatric issues than others with an IDD	Rare, unusual occurrences of significant slowing and regression, but not dementia, in young adults
				About age 51–55 is the mean age of onset for Alzheimer's *disease*

From Bull, M.J., Trotter, T., Santoro, S.L., Christensen, C., Grout, R.W., Council on Genetics, 2022. Health supervision for children and adolescents with down syndrome. Pediatrics 149(5), e2022057010MEDICAL; Charlot, L., Fox, S., Friedlander, R., 2002. Obsessional slowness in down's syndrome. J. Intellect. Disabil. Res. 46 (6), 517–524.

are affected by developmental problem areas such as communication and social awareness. Epigenetic factors may impact the nature and extent of an individual's developmental challenges and thus contribute to aggression risk in this way.

It is important to remember that ACEs, which can result in aggressive and violent behavior in adults, also operate through epigenetic mechanisms (Muller & Kenney, 2024).

In summary, epigenetic dysregulation is a convergent pathway in the pathogenesis of disorders underlying ASD/IDD, as well as behavioral patterns characterized by aggression, integrating genetic and environmental risk factors and offering potential targets for future therapies and biomarkers.

Summary

The need to know as much as possible about a person who has ASD or another IDD and aggressive behavior includes a comprehensive understanding of their history from prenatal life through childhood to the present, including information on family history, the mother's pregnancy, early childhood experiences, growth patterns, physiology and medical conditions, social and environmental circumstances, and genetic profile. Medical conditions associated with genetic syndromes often contribute to the risk of behavioral manifestations such as anxiety, distress, and aggression. The burgeoning field of epigenetics and new ways of understanding how gene-environment interactions affect function and behavior point the way to a future with potential interventions to prevent behavioral challenges or even prevent the occurrence of some NDDs.

It may seem counterintuitive, but high heritability has little bearing on whether specific interventions may alter individual outcomes. This is important to understand for people with ASD/IDD in general, but specifically in relation to how these individuals develop patterns of behavior, such as aggressive acts.

It is best to think of outcomes as probabilistic. This means we focus on there being an increased likelihood of certain behaviors or concerns resulting from genotypes and descriptions we receive of typical phenotypes. Unique versus shared behaviors are also of interest: "not every person with a specific genetic disorder necessarily shows that disorder's characteristic features" (Rutter 1978). Be careful not to expect that every person with a given syndrome will exhibit all the major features, or assume that someone without *all* the described features cannot have the syndrome. It is therefore important to take a thorough history of early childhood experiences among individuals with ASD/IDD who have significant problems with aggression.

As discussed in Chapter 2, developmental and neurodevelopmental strengths and deficits are shaped in multiple ways and affect how people respond to stress. Essentially, a greater degree of global neurodevelopmental burden adds risk for aggressive behavior. How caregivers and other environmental factors respond to a person experiencing difficulty further determines outcomes. The probability of aggressive behavior may increase when environmental agents are overwhelmed and correspondingly decrease when environments respond in a supportive manner. The use of a combined developmental and contextual approach is described in Chapter 9.

References

Álvarez-Mora, M.I., Sánchez, A., Rodríguez-Revenga, L., Corominas, J., Rabionet, R., Puig, S., Madrigal, I., 2022. Diagnostic yield of next-generation sequencing in 87 families with neurodevelopmental disorders. Orphanet J. Rare Dis. 17 (1), 60.

ASD. (2025). The genetic mystery of why some people develop autism. BBC Future. https://www.bbc.com/future/article/20250415-the-genetic-mystery-of-why-some-people-develop-autism

Balachandar, V., Mahalaxmi, I., Neethu, R., Arul, N., Abhilash, V.G., 2022. New insights into epigenetics as an influencer: an associative study between maternal prenatal factors in autism spectrum disorder (ASD). Neurol. Perspect. 2667-0496. 2 (2), 78–86. doi:10.1016/j.neurop.2022.01.002.

Birnbaum, R., Mahjani, B., Loos, R.J.F., Sharp, A.J., 2022. Clinical characterization of copy number variants associated with neurodevelopmental disorders in a large-scale multiancestry biobank. JAMA Psychiatry. 2168-622X. 79 (3), 250–259. Available from: http://archpsyc.jamanetwork.com/issues.aspx. doi:10.1001/jamapsychiatry.2021.4080.

Burke, C., Ellis, J.D., Peltier, M.R., Roberts, W., Verplaetse, T.L., Phillips, S., Moore, K.E., Marotta, P.L., McKee, S.A., 2023. Adverse childhood experiences and pathways to violent behavior for women and men. J. Interpers. Violence. 1552-6518. 38 (3–4), 4034–4060. Available from: https://journals.sagepub.com/home/JIV. doi:10.1177/08862605221113012.

Chaves, T.F., Ocampos, M., Barbato, I.T., de Camargo Pinto, L.L., de Luca, G.R., Filho, J.H.B., Bernardi, P., Muniz, Y.C.N., Maris, A.F., 2024. A cohort study of neurodevelopmental disorders and/or congenital anomalies using high resolution chromosomal microarrays in southern Brazil highlighting the significance of ASD. Sci. Rep. 2045-2322. 14 (1). Available from: https://www.nature.com/srep/. doi:10.1038/s41598-024-54385-2.

Cox, D., The genetic mystery of why some people develop autism. BBC 16 April 2025 (n.d.) Available from: https://www.bbc.com/future/article/20250415-the-genetic-mystery-of-why-some-people-develop-autism.

Dall'Aglio, L., Muka, T., Cecil, C.A.M., Bramer, W.M., Verbiest, M.M.P.J., Nano, J., Hidalgo, A.C., Franco, O.H., Tiemeier, H., 2018. The role of epigenetic modifications in neurodevelopmental disorders: a systematic review. Neurosci. Biobehav. Rev. 1873-7528. 94, 17–30. Available from: www.elsevier.com/locate/neubiorev. doi:10.1016/j.neubiorev.2018.07.011.

Doi, M., Usui, N., Shimada, S., Prenatal environment and neurodevelopmental disorders. Front. Endocrinol. (Lausanne). 1664-2392. 2022. 13, 860110. Available from: https://www.frontiersin.org/journals/endocrinology#. doi:10.3389/fendo.2022.860110. PMID: 35370942; PMCID: PMC8964779.

Fortea, J., Zaman, S.H., Hartley, S., Rafii, M.S., Head, E., Carmona-Iragui, M., 2021. Alzheimer's disease associated with Down syndrome: a genetic form of dementia. Lancet Neurol. 1474-4465. 20 (11), 930–942. Available from: http://www.journals.elsevier.com/the-lancet-neurology/. doi:10.1016/S1474-4422(21)00245-3.

Herrera, M.L., Paraíso-Luna, J., Bustos-Martínez, I., Barco, Á., 2024. Targeting epigenetic dysregulation in autism spectrum disorders. Trends Mol. Med. 1471-499X. 30 (11), 1028–1046. Available from: https://www.sciencedirect.com/science/journal/14714914. doi:10.1016/j.molmed.2024.06.004.

Karam, S.M., Riegel, M., Segal, S.L., Félix, T.M., Barros, A.J.D., Santos, I.S., Matijasevich, A., Giugliani, R., Black, M., 2015. Genetic causes of intellectual disability in a birth cohort: a population-based study. Am. J. Med. Genet. A. 1552-4833. 167 (6), 1204–1214. Available from: http://onlinelibrary.wiley.com/journal/10.1002/(ISSN)1552-4833. doi:10.1002/ajmg.a.37011.

Khan, A., Barapatre, A.R., Babar, N., Doshi, J., Ghaly, M., Patel, K.G., Nawaz, S., Hasana, U., Khatri, S.P., Pathange, S., Pesaru, A.R., Puvvada, C.S., Billoo, M., Jamil, U., 2025. Genomic medicine and personalized treatment: a narrative review. Ann. Med. Surg. 87 (3), 1–8. doi:10.1097/MS9.0000000000002965.

Klahr, A.M., Thomas, K.M., Hopwood, C.J., Klump, K.L., Burt, S.A., 2013. Evocative gene–environment correlation in the mother–child relationship: a twin study of interpersonal processes. Dev. Psychopathol. 25 (4 Pt 1), 1051–1068. doi:10.1017/S0954579413000372.

La Monica, I., Di Iorio, M.R., Sica, A., Rufino, F., Sotira, C., Pastore, L., Lombardo, B., 2025. Autism Spectrum Disorder: genetic Mechanisms and Inheritance Patterns. Genes. 2073-4425. 16 (5), 478. doi:10.3390/genes16050478.

LaSalle, J.M., 2023. Epigenomic signatures reveal mechanistic clues and predictive markers for autism spectrum disorder. Mol. Psychiatry. 1476-5578. 28 (5), 1890–1901. Available from: https://www.nature.com/mp/. doi:10.1038/s41380-022-01917-9.

Lenz, S., Sivaloganathan, A., Goodman, S.J., Cytrynbaum, C., Rapley, J., Canning, E., Baribeau, D., 2025. Psychopharmacology in children with genetic disorders of epigenetic and chromatin regulation. J. Neurodev. Disord. 1866-1955. 17 (1). Available from: https://jneurodevdisorders.biomedcentral.com/. doi:10.1186/s11689-025-09605-9.

Lopes-Marques, M., Mort, M., Carneiro, J., Azevedo, A., Amaro, A.P., Cooper, D.N., Azevedo, L., 2024. Meta-analysis of 46,000 germline de novo mutations linked to human inherited disease. Hum. Genomics. 1479-7364. 18 (1). Available from: https://humgenomics.biomedcentral.com/. doi:10.1186/s40246-024-00587-8.

Love, C., Sominsky, L., O'Hely, M., et al, 2024. Prenatal environmental risk factors for autism spectrum disorder and their potential mechanisms. BMC Med. 22 (393). doi:10.1186/s12916-024-03617-3.

Millan, M.J., 2013. An epigenetic framework for neurodevelopmental disorders: from pathogenesis to potential therapy. Neuropharmacology 68, 2–82.

Müller, R., Kenney, M., 2024. The evolution of ACEs: from coping behaviors to epigenetics as explanatory frameworks for the biology of adverse childhood experiences. Hist. Philos. Life Sci. 46(4), 33. doi:10.1007/s40656-024-00629-3. PMID: 39470844; PMCID: PMC11522112.

Rangasamy, S., D'Mello, S.R., Narayanan, V., 2013. Epigenetics, Autism Spectrum, and Neurodevelopmental Disorders. Neurotherapeutics. 1878-7479. 10 (4), 742–756. doi:10.1007/s13311-013-0227-0.

Reichard, J., Zimmer-Bensch, G., 2021. The epigenome in neurodevelopmental disorders. Front. Neurosci. 1662-453X. 15. doi:10.3389/fnins.2021.776809.

Rodan, L.H., Stoler, J., Chen, E., Geleske, T., Vergano, S., Holm, I.A., Jones, K., Kalish, J., Monteil, D.C., Pritchard, A.B., Rasmussen, S.A., Russell, B., Santoro, S.L., Trapane, P., Weaver, K.N., 2025. Genetic evaluation of the child with intellectual disability or global developmental delay: clinical report. Pediatrics. 1098-4275. 156 (1). Available from: https://watermark.silverchair.com. doi:10.1542/peds.2025-072219.

Rutter, M., 1978. Social class, genetic factors, and cognitive development. In: Jensen, A.R., Ed., Genetics and Education. Methuen, London, pp. 59–88.

Sanchis-Juan, A., Megy, K., Stephens, J., Ricaurte, C.A., Dewhurst, E., Low, K., French, C.E., Grozeva, D., Stirrups, K., Erwood, M., McTague, A., Penkett, C.J., Shamardina, O., Tuna, S., Daugherty, L.C., Gleadall, N., Duarte, S.T., Hedrera-Fernández, A., Vogt, J., Ambegaonkar, G., Chitre, M., Josifova, D., Kurian, M.A., Parker, A., Rankin, J., Reid, E., Wakeling, E., Wassmer, E., Woods, C.G., Raymond, F.L., Carss, K.J., 2023. Genome sequencing and comprehensive rare-variant analysis of 465 families with neurodevelopmental disorders. Am. J. Hum. Genet. 1537-6605. 110 (8), 1343–1355. Available from: http://www.elsevier.com/wps/find/journaldescription.cws_home/713561/description#description. doi:10.1016/j.ajhg.2023.07.007.

Sandoval-Talamantes, A.K., Mori, M.Á., Santos-Simarro, F., García-Miñaur, S., Mansilla, E., Tenorio, J.A., Peña, C., Adan, C., Fernández-Elvira, M., Rueda, I., Lapunzina, P., Nevado, J., 2023. Chromosomal microarray in patients with non-syndromic autism spectrum disorders in the clinical routine of a tertiary hospital. Genes. 2073-4425. 14 (4). Available from: http://www.mdpi.com/journal/genes/. doi:10.3390/genes14040820.

Schuurmans, I.K., Smajlagic, D., Baltramonaityte, V., Malmberg, A.L.K., Neumann, A., Creasey, N., Felix, J.F., Tiemeier, H., Pingault, J.B., Czamara, D., Raïkkönen, K., Page, C.M., Lyle, R., Havdahl, A., Lahti, J., Walton, E., Bekkhus, M., Cecil, C.A.M., Genetic susceptibility to neurodevelopmental conditions associates with neonatal DNA methylation patterns in the general population an individual participant data meta-analysis. SSRN, Netherlands SSRN. 1556-5068. (2024), Available from: https://www.ssrn.com/index.cfm/en/. doi: 10.2139/ssrn.4882699.

Shin, S., Lee, J., Kim, Y.G., Ha, C., Park, J.H., Kim, J.W., Lee, J., Jang, J.H., 2023. Genetic Diagnosis of Children With Neurodevelopmental Disorders Using Whole Genome Sequencing. Pediatr. Neurol. 1873-5150. 149, 44–52. Available from: www.elsevier.com/locate/pedneu. doi:10.1016/j.pediatrneurol.2023.09.003.

Shook, L.L., Castro,V., Ibanez-Pintor, L., Perlis, R.H., Edlow,A.G., 2025. Neurodevelopmental outcomes of 3-year-old children exposed to maternal severe acute respiratory syndrome Coronavirus 2 (SARS-CoV-2) infection in utero. Obstet. Gynecol. 0029-7844. doi:10.1097/aog.0000000000006112.

Stenseth, N.C., Andersson, L., Hoekstra, H.E., 2022. Gregor Johann Mendel and the development of modern evolutionary biology. Proc. Natl. Acad. Sci. U. S. A. 119(30), e220132719. doi:10.1073/pnas.2201327119. Epub 2022 Jul 18. PMID: 35858454; PMCID: PMC9335310.

Tseng, C.E.J., McDougle, C.J., Hooker, J.M., Zürcher, N.R., 2022. Epigenetics of autism spectrum disorder: histone deacetylases. Biol. Psychiatry. 1873-2402. 91 (11), 922–933. Available from: www.elsevier.com/locate/biopsychiat. doi:10.1016/j.biopsych.2021.11.021.

van der Sanden, B.P.G.H., Schobers, G., Galbany, J.C., Koolen, D.A., Sinnema, M., van Reeuwijk, J., Stumpel, C.T.R.M., Kleefstra, T., de Vries, B.B.A., Ruiterkamp-Versteeg, M., Leijsten, N., Kwint, M., Derks, R., Swinkels, H., den Ouden, A., Pfundt, R., Rinne, T., de Leeuw, N., Stegmann, A.P., Stevens, S.J., van den Wijngaard, A., Brunner, H.G., Yntema, H.G., Gilissen, C., Nelen, M.R., Vissers, L.E.L.M., 2023. The performance of

genome sequencing as a first-tier test for neurodevelopmental disorders. Eur. J. Hum. Genet. 1476-5438. 31 (1), 81–88. Available from: https://www.nature.com/ejhg/. doi:10.1038/s41431-022-01185-9.
Vissers, L.E., Gilissen, C., Veltman, J.A., 2015. Genetic studies in intellectual disability and related disorders. Nat. Rev. Genet. 16 (1), 9–21.
Wang, J., Lan, Y., Qi, H.Y., Wang, L.H., Wei, S., Yuan, Y., Luo, T., 2025. Comparison of fluorescence in situ hybridization, next-generation sequencing, and DNA methylation microarray for copy number variation assessment in gliomas. Lab. Investig. 105(7), 104168.

Further reading

Dimachkie Nunnally, A., Nguyen, V., Anglo, C., Sterling, A., Edgin, J., Sherman, S., Thurman, A.J., 2021. Symptoms of autism spectrum disorder in individuals with Down syndrome. Brain Sci. 11 (10), 1278.
Dorsett, D., Fay, J. (2013). U.S. Patent Application 13/889,314. http://ghr.nlm.nih.gov/gene/NIPBL
Glasson, E.J., Buckley, N., Chen, W., Leonard, H., Epstein, A., Skoss, R., Downs, J., 2020. Systematic review and meta-analysis: mental health in children with neurogenetic disorders associated with intellectual disability. J. Am. Acad. Child Adolesc. Psychiatry 59 (9), 1036–1048.
Gruca-Stryjak, K., Doda-Nowak, E., Dzierla, J., Wróbel, K., Szymankiewicz-Bręborowicz, M., Mazela, J., 2024. Advancing the clinical and molecular understanding of Cornelia de Lange syndrome: a multidisciplinary pediatric case series and review of the literature. J. Clin. Med. 13 (8), 2423.
Josselsohn, A., Zhao, Y., Espinoza, D., Hollander, E., 2024. Oxytocin in neurodevelopmental disorders: autism spectrum disorder and Prader-Willi syndrome. Pharmacol. Ther. 264, 108734.
Kazdoba, T.M., Leach, P.T., Crawley, J.N., 2016. Behavioral phenotypes of genetic mouse models of autism. Genes Brain Behav. 15 (1), 7–26.
Larsen, F.K., Baksh, R.A., McGlinchey, E., Langballe, E.M., Benejam, B., Beresford-Webb, J., McCarron, M., Coppus, A., Falquero, S., Fortea, J., Levin, J., Loosli, S.V., Mark, R., Rebillat, A., Zaman, S., Strydom, A., 2024. Age of alzheimer's disease diagnosis in people with down syndrome and associated factors: results from the horizon 21 European down syndrome consortium. Alzheimers Dement. 1552-5260. 20 (5), 3270–3280. doi:10.1002/alz.13779.
Metzler, S., Brown, G.R., 2025. A review of Prader-Willi syndrome. JAAPA 38 (2), e1–e6.
van der Lei, M.B., Kooy, R.F., 2025. From discovery to innovative translational approaches in 80 years of Fragile X syndrome research. Biomedicines 13 (4), 805.

CHAPTER 5

Psychiatric disorders and risk for aggressive behavior

Lauren Charlot[a] and Angela Hassiotis[b]
[a]UMass Chan Medical School, Worcester, MA, United States
[b]University College London, London, United Kingdom

Individuals with autism spectrum disorder (ASD) or other intellectual and developmental disabilities (IDD) experience higher rates of psychiatric conditions and behavioral challenges compared to people without neurodevelopmental disabilities (Cooper et al., 2007; Lunsky et al., 2018). Prevalence estimates vary based on setting, methods, and diagnostic tools (Esler et al., 2019). Using the Diagnostic and Statistical Manual of Mental Disorders, Fifth Edition (DSM-5), and former DSM criteria, rates of psychiatric comorbidity have been reported to be as high as 70% among people with ASD (Edition, 2013; Magyar and Pandolfi, 2017). Some systems distinguish psychiatric syndrome diagnoses from diagnoses of behavioral disorders. Studies of children with ASD that consider the likely impact of developmental effects on presentations of mental illness report different findings compared to retrospective chart review studies (Leyfer et al., 2006; Mazefsky et al., 2012; Magyar and Pandolfi, 2017). In some studies, the specific diagnoses assigned varied depending on whether the assessment setting was specialized for the care of patients with ASD/IDD or was a typical or "generic" psychiatric setting. Lunsky and Bradley (2007) found that people with IDD in specialized inpatient settings were diagnosed with anxiety and depression more often, while those in more generic settings were diagnosed more frequently with psychosis.

People with autism or other IDD referred for psychiatric assessment often present with some form of aggressive behavior (Smith et al., 2022; Tsiouris et al., 2011). This has provoked debates as to the meaning of behaviors that challenge and of aggression, in particular, in the context of possible psychiatric illness (Allan, 2008; Charlot et al., 1993). Aggression and irritability have been identified as occurring in conjunction with many different psychiatric disorders and with other sources of stress in studies of people with IDD (Charlot et al., 1993; Tsiouris

Aggressive Behavior of People with Autism and other Intellectual and Developmental Disabilities
https://doi.org/10.1016/B978-0-443-26552-5.00002-3

et al., 2011; Buie et al., 2010; Crocker et al., 2014; Smith et al., 2022). Aggression and other behaviors that challenge indicate something is wrong but have limited specificity as to what the root causes may be for any given patient (Charlot et al., 1993; Hunter et al., 2008). Externalizing behaviors may be seen as a manifestation of distress in a person with few ways to express this (due to limited language or cognitive challenges). As a result, *Best Practice* guidelines urge the use of a multidisciplinary assessment when psychotropics are prescribed primarily to treat aggression (Bealsey et al., 2021; Charlot et al., 2019; National Institute for Health and Care Excellence, 2015). Multidisciplinary assessment may also be important when usual psychiatric care has been ineffective, as incorrect psychiatric labels may lead to ineffective treatment choices and even some that cause yet more aggressive behavior.

Many reports identify an overreliance on psychotropic medications and polypharmacy used to treat emotional and behavioral challenges in the population, with significant use of antipsychotics to reduce aggression (Branford et al., 2019; Esler et al., 2019; Ritter et al., 2021; Scheifes et al., 2016). Polypharmacy, in turn, may be associated with added medical problems that provoke or worsen aggressive behaviors (Charlot et al., 2011; Hassiotis and Rudra, 2022; O'Brien et al., 2024). In addition, people with IDD appear to suffer high rates of medication side effects, in some cases more so than individuals without a neurodevelopmental disability (Charlot et al., 2020; Sheehan and Hassiotis, 2017). High rates of emergency psychiatric service use have been reported among a subset of the population (Beasley et al., 2018; Jeglum et al., 2024; Lindgren et al., 2021; Hassiotis and Vegh, 2024). It is unclear if individuals experiencing a high need for emergency psychiatric services have inaccurate diagnoses, and more examination of these questions is warranted. Many individuals referred to emergency care were aggressive, prompting reliance on medical emergency services or stays in an emergency department (ED) (Beasley et al., 2018). In all of the above instances, assessment to determine if an autistic person or a person with ID who displays aggressive behavior has a specific psychiatric diagnosis requires consideration of the impact of many variables on manifestations.

Individuals with autism and other IDD are frequently found to suffer from medical and neurological comorbidities that complicate psychiatric differential diagnostic assessments, especially the assessment of the roots of aggressive behavior. The most common medical comorbidities are discussed in Chapter 3. Getting the wrong psychiatric diagnosis likely leads to ineffective treatment but may also intensify or provoke aggressive behavior as a result. Despite these concerns, few studies address the extent to which people with autism or other IDD may have inaccurate psychiatric diagnoses (Charlot et al., 2022).

Psychiatric assessment of people with ASD/IDD

Psychiatric assessment involves significant input from the person being assessed, even when they have difficulties expressing themselves. In addition, informants will be interviewed, including some paid caregivers, family members, and professionals available to assist in the assessment. Different informants will have varied points of view. As noted by Hassiotis and Vegh (2024) the preferred approach is one in which a team works together to perform a comprehensive assessment. Tools for informant interviews that will produce a suspected diagnosis from DSM-5 and ICD-11 are available but limited in number. The MOSS-ADD and the MASS are two tools that yield possible DSM or ICD diagnoses. Other instruments may be useful to determine the efficacy of treatments for irritability and aggression but are not specifically aimed at producing suspected psychiatric diagnoses. Yet other tools are structured and checklist-style, without interviews, but do point to suspected psychiatric diagnoses. For the most accurate diagnosis, we recommend assessment be multidisciplinary, and if instruments are used, they should be ones with some flexibility (semistructured interviews) tied to DSM or ICD. In Chapter II, the ways in which a developmental framework enhances assessment and treatment are reviewed in detail. The importance of developing a case formulation is discussed next. In an effort to develop a case formulation when the display of aggressive behavior is a key component bringing a person to clinical attention, it is critical to consider the contribution of developmental features. While the importance of a biopsychosocial approach is well recognized, the addition of a developmental perspective is urged here. For example, recognizing what is or is not a psychotic symptom involves attention to the person's often developmentally driven means of processing information and describing their thoughts to others. In general, psychiatric phenomena should represent something that is a "departure from usual behavior" for the individual. It is important not to mistake misunderstanding for a lack of reality-based conceptualizations (Box 5.1). As suggested by Solomon:

Box 5.1 Mental status exam for an interview of patient with ASD/IDD

Appearance:

Observations of how the person looks.

- Is he or she neatly dressed? Well groomed? (hair brushed or combed) or appears disheveled?
- Dressed for the season or weather—not, i.e., dressed for winter but it is summer?
- Tall, short-average build? Obese, thin-appearing?
- Are there any dysmorphic features?
- Head circumference, posturing, i.e., dystonic postures, spasticity.

Level of alertness: Is the patient awake, alert, and conscious?

If not, can they be aroused?

- Can the person focus on your questions and conversation? What is their attention span? (Critical to this—is this usual for this person based on their baseline when they are well?)
- Does he or she seem to connect with you for 1 min, and then nod off or become distant?
- Is there a "waxing and waning" of alertness?
- A "clouded sensorium" seems foggy ... in and out of it.

Motor behavior

Any atypical or abnormal movements?

- Tics, twitches
- Tremors
- Ataxia
- Toe walking
- Stereotypies such as hand-waving and rocking
- Rapid eye blinking
- Oral buccal abnormal movements (dyskinesias)
- Leaning to one side, unbalanced
- Unable to stay seated
- Overactive-pacing, restless fidgety movements
- Hand wringing

Speech: Is it normal in tone, volume, rate, etc.?

- *Volume*—Soft spoken, loud
- *Quantity* (based on what you understand about the person's profile)? Says very little; gives long answers.
- *Quality*—Speech can be monotone, bulleted, or have an odd prosody (sing-song), with single words or phrases.
- *Rate* may be rapid (talks very fast), with words running together.
- Difficult to understand due to dysarthria (garbled speech), slurred speech, with echolalia or stuttering.

BOX 5.1 Mental status exam for an interview of patient with ASD/IDD—cont'd

- *Content*—is it goal-directed? Off-topic? Unfocused? (How does this align with the baseline?)
- Long latency may be observed,

Behavior with regard to the interview:

- Is he or she guarded, looking tense about the interview, or cooperative and pleasant?
- Is this what would be expected for this person under usual circumstances?

Orientation:

- Is he or she oriented to person, place, and time, as well as possible for their usual abilities?
- Does the person know where they are and what they are doing here? Can this person explain why they are having an assessment and say what, if anything, has been troubling them?
- Does he or she know who the people around them are?
- Can the person tell you the day, date, and year? Remember the clinician's name?
- Most important—Is all of this reflecting what is "normal" for this person when well?

Mood: How do they feel? (Ask this directly if the person is at all able to answer; use visuals.)

- *How have you been feeling lately?*
- *Are you happy?*
- *Are you sad?*
- *Have you been feeling depressed?*
- *Are you worried or nervous? What worries you?*
- *Are you angry? Are you usually not mad or angry with anyone*?
- Are the answers appropriate for their current situation?
- Does he or she usually say yes to any question? Reverse the questions: does he or she say, "Yes, I am sad," and agree that, "I have not been sad"?
- Is the answer usual for the person? If he or she cannot answer, ask the informants for signs of these various mood states:
  i. Crying, smiling, tense facial expressions, restlessness, worrying, looking scared, upset, seeming angry (shaking fist at others and scowling), showing a loss of interest in things he or she used to enjoy, unable to seem to have fun anymore, or be happy.
  ii. If aggression is reported as a "mood" sign or symptom, then obtain a detailed description of this. Informants may focus on externalizing behaviors and make assumptions about underlying affect, i.e., report that the person is angry, irritable, manic, or psychotic when the person is actually anxious, frustrated, disappointed, or not revealing their mood state.

Continued

BOX 5.1 Mental status exam for an interview of patient with ASD/IDD—cont'd

Affect: How does the patient appear to you? This interpretation is based on your observation of their interactions during the interview.

- Does he or she make eye contact? Is that usual for him or her when well?
- Is he or she over active or very excited in response to your interaction?
- Common assessments include: person is
  - anxious (tense appearance, fidgeting, hand-wringing, furtive looks, etc.);
  - flat (unchanging throughout— isn't smiling when usually someone would, or crying or looking sad when discussing sad things);
  - excitable;
  - angry;
  - sad, downcast; and
  - appropriate.

Thought process: This is a description of the way in which a person thinks and processes information. This really cannot be assessed in people who cannot clearly verbalize, and without information about cognition when well (including executive functions such as short-term memory, processing speed, mental flexibility, or capacity to shift response sets, attention, and understanding of serial commands or multistep instructions).

- Are comments logically organized? If not, how off-base are they?
- Does the person tend to stray off-topic? Is this new?
- Are thoughts appropriately linked, or are they simply all over the map? Is this new?
- Most important—is this usual? (For many people with IDD, this is usual.)

Thought content: A description of what the patient is thinking about and whether their thoughts appear logical and reality-based.

To assess this too, you must know this person's cognitive capacities when they are well. The point is to identify something now that is out of line with what is typical for this person, i.e., usually understands the days of the week—right now, seems confused about what day it is.

Or really thinks the characters in movies are real, so talking about superheroes when nervous is usual. Some people with IDD at baseline will run from one topic to another, and this is because they lack the cognitive capacities to comprehend the logical relations among topics. This should not be mistaken for the acute appearance of illogical thinking or tangential thinking seen in psychosis or mania.

Based on history

- Is he or she able to completely understand reality versus fantasy?
- More of a concrete thinker (based on what you can see, touch, and count).
- Understands time concepts: hours, days, weeks, years?
- Able to think abstractly—have theoretical plans or ideas.
- Able to have hopes for the future (is he or she hopeful?)

BOX 5.1 Mental status exam for an interview of patient with ASD/IDD—cont'd

MSE: If able to discuss, is he or she worried or anxious about things? (If yes, what about?)

- Afraid bad things will happen?
- What are the fears?
- Are these fears out of proportion to real concerns?
- Are these fears linked to a history of traumatic events?
- Fears related to not being the one who decides what will happen?

NOTE: The presence of psychotic thinking can be extremely difficult to establish unless the person has a mild cognitive impairment, is articulate, and you know his or her baseline in terms of *executive functions*. See (Dossetor, 2007).

- A person cannot have paranoid thinking if he or she does not also have a clear understanding that others have different perspectives (*Theory of Mind*, or awareness of competing perspectives). There is a need to understand that people are able to plan things (future orientation).
- Fears of bad things or bad actions aimed at a person are often grounded in real histories of trauma and negative experiences. Limits to internal processing cause this to seem psychotic when it is simply a lack of understanding.

Delusions (i.e., holding beliefs that are untrue) can only be established if the person's unique perspectives on "reality" are known, as many people with IDD will confuse fantasy-reality concepts, express wishes as facts, and simply not understand when a story seems odd or unreal. They may confabulate rather than lie (more as a younger neurotypical child might prior to about age 6–7). People with ASD may have what seems to others as bizarre ideas related to restrictive and repetitive interests or behaviors. Some people state private wishes out loud, not knowing that it looks odd. The presence of *auditory hallucinations* can be difficult to determine when some people with even mild ID cannot differentiate their own thoughts and reflections from a "voice" in their head.

People with IDD may also be more prone to simply trying to please the examiner and giving the affirmative answer that it seems the examiner wants. Many people will redo conversations with themselves internally and report hearing voices.

Suicidal ideation—clinicians should be cautious not to word all their questions in a way that can be answered with just a yes or no, and they should be sure to ask about a *plan* if one is suspected to be present, as well as assess the opportunity for self-harm (i.e., a person on 1:1 supervision 24/7).

Some people with IDD have learned that saying they are suicidal will get clinicians to provide more support. They may need help just to know it is okay to *ask for help*. However, suicidality is real for people with IDD too and should not be overlooked.

Intent to harm self or others—in some cases, may be revealed through recent and historical actions even more so than through an interview; both should be conducted.

Continued

BOX 5.1 Mental status exam for an interview of patient with ASD/IDD—cont'd

If there is any concern regarding possible suicide or homicide, the patient should be asked about this directly, and details should be sought (e.g., specific plan, time, etc.).

Memory*: Short-term memory* is usually assessed by listing three objects, asking the patient to repeat them to you to ensure that they were heard correctly, and then checking recall at 5 min.

- For people with IDD, again—must know baseline.
- Is he or she forgetting things that are usually easy for him or her to remember?
- At baseline, how many steps can the person follow?
- If possible, review psychological testing that reports on Working Memory and Processing Speed, both of which may be lower than IQ and will impact the capacity for self-control and mood regulation.
- For people with IDD, even with very mild ID—short-term or **working memory** challenges may be very significant at baseline.
  - Working memory problems may indicate difficulty storing information.
  - Short-term—may represent difficulty with retrieval, sometimes related to anxiety.

Long-term memory can be evaluated by asking about history, such as where one was born and raised, family history, but it will, as with other areas, be impacted by baseline abilities to reliably describe one's own history.

Ability to perform calculations: Can the person perform simple addition and multiplication? Are the responses in line with baseline expectations?

If able, digit span can be performed. Is recall okay forward, but not backward?

Judgment: Provide a common scenario and ask, "What would you do?" (e.g., "If you found a letter on the ground in front of a mailbox, what would you do with it?").

Must know the person's baseline to determine if what is reported now represents a departure from their usual behavior, and make note of this. If, at their best, the person would not be able to judge safety or socially expected responses, this is, of course, not a measure of an emerging symptom.

Executive functioning, information processing, and ***reasoning***: Involves the interpretation of complex ideas.

i.e., one may ask the meaning of a proverb, "People in glass houses should not throw stones." A few common interpretations include: concrete (e.g., "Don't throw stones because it will break the glass"); abstract (e.g., "Don't judge others"). Concrete answers are very common at baseline for many people with IDD and are not a reflection of an acute process.

Certain psychiatric, neurological, and other medical conditions are characterized by patterns that can be identified through the analysis of the answers to the MSE questions when this information is combined with an interview of a familiar informant and a review of available records.

BOX 5.1 Mental status exam for an interview of patient with ASD/IDD—cont'd

Added review items: This has to be determined by asking the person or their caregivers:

Sleep pattern (24 h)

Energy level

Activity engagement

Eating and drinking—any problems and/or any alterations that have cooccurred with symptoms leading to this assessment?

Bowel and urination patterns—any problems and/or alterations that have cooccurred with symptoms leading to this assessment.

Medical problems: Note any recent changes in these, as at times these are missed causes of irritability and agitation, or they lower the threshold for outbursts.

Medication and suspected side effects: Note any changes in these, as at times these are missed causes of irritability and agitation or may lower the threshold for outbursts. Especially common adverse drug events are constipation, sedation, and increased appetite.

When Formal Thought Disorder is present in ASDs, it generally is not a co-morbid schizophrenia symptom, but is related to pragmatic language abnormalities found in ASDs.

(Solomon et al., 2008, p. 1474)

Psychiatric diagnoses and aggressive behavior: Developmental influences

It is important to recognize the interactions between a person's unique profile of neurodevelopmental skills and challenges and the contexts in which they have lived. An example is when a person with autism lacks an understanding of competing perspectives and inadvertently upsets school peers by striking out when anxious, then fails to comprehend their rejection. In some cases, this is seen as antisocial behavior but may actually be an example of failing to understand typical social expectations.

A special challenge in the United States is the lack of training for psychiatrists and other mental health providers in the special issues and skills that facilitate specialized care for people with ASD/IDD who have behavioral health needs (Beasley et al., 2021). One of the most common confusions that arises in the process of diagnosing a mental health condition in a person with a neurodevelopmental disorder is establishing what is an actual departure from usual behavior for that person. This should be a consideration in any psychiatric assessment. However, we are sometimes assessing adults who are processing information in ways we are far more accustomed to seeing in a child. When an adult believes movie characters are real, is he exhibiting usual behavior or is this pathological—even delusional? Adults with a neurotypical developmental history might respond one way to difficulties, while people with an atypical profile related to neurodevelopmental factors may respond quite differently. It is understood that youth present things like sadness, grief, and anxiety differently than adults do when all have a neurotypical developmental history. There may be assumptions about what are expected responses to stressful life events based on our understanding of usual behavior for people of a certain age or from a specific culture. A similar framework is needed to assess observations and reports when the patient has neurodevelopmental challenges that impact information processing and social awareness. Incorrect psychiatric diagnoses may stem from reports of symptoms that are misunderstood.

In some cases, the intensity of a stress response may be impacted by developmental challenges that affect a person's capacity to manage stress generally. For example, self-control is, in part, a function of the maturity of certain frontal lobe functions, or a person's *executive functions* (see Chapter 2). If, related to a neurodevelopmental condition, an individual is unable to reason and then refrain from a behavioral response to a stressor, is this a sign of mental illness? Is it a sign of acute mental illness when a person reacts loudly and negatively to something they dislike rather than asking to be excused from it or planning a way to manage the stress? In many instances, people with ASD or other IDD are *escorted* to face things they find stressful without being asked about their preferences. Is it a delusion or a wish when a person interacts with a kindly grocery store check-out staff member and then describes her as a potential wife?

It is also helpful to consider what constitutes a stressful life event for a person based on *their* unique profile and history. For example, some individuals with autism or other IDD have a history of rejection in social situations related to their lack of social awareness and "usual" social expectations. Over time, being pressured to be in social situations becomes aversive. Yet, this individual may live in a group residence as an adult and be required to repeatedly attend group activities.

Some assessment tools that can be used with direct evaluations are shown in Box 5.2.

Box 5.2 Examples of diagnostic assessment interview tools that yield suspected DSM or ICD psychiatric diagnoses

ACI-PL

The ACI-PL, a structured-interview instrument designed to diagnose psychiatric disorders in children and adolescents with ASD, was piloted, and frequently diagnosed disorders—depression, ADHD, and OCD—were tested for reliability and validity. The tool was modified from the Kiddie Schedule for Affective Disorders and Schizophrenia.

The ACI-PL provides reliable DSM diagnoses that are valid, based on clinical psychiatric diagnosis and treatment history. The sample demonstrated a high prevalence of specific phobia, obsessive-compulsive disorder, and ADHD. (See https://psycnet.apa.org/record/2006-13225-003.)

MASS

The MASS (Mood and Anxiety Semi-Structured) Interview is an informant interview tool used in three published reports. To establish validity, the MASS Interview-derived diagnoses were compared with clinical DSM-IV diagnoses derived from an extensive inpatient evaluation and classifications based on the Hamilton Depression Rating Scale for 93 psychiatric inpatients served on a specialized unit for people with ASD/IDD and major mental health disorders. Agreement with the MASS Interview was high, yielding significant kappa coefficients ranging from 0.42 to 0.78. (Charlot (2007)—Wiley Online Library; Charlot et al., 2022; Charlot et al., 2008.)

Moss Psychiatric Assessment Schedule (Moss-PAS)

The Moss-PAS is an updated version of the PAS-ADD, featuring a suite of assessments that includes clinical interviews and a checklist (see below).

The MOSS-PAS (ID) is designed to provide highly reliable information on psychiatric symptoms in adults, usually through an informant interview. It is not a screening instrument but a sophisticated assessment that uses a glossary of symptom definitions to guide the coding.

The Moss-PAS (ID) can be used to collect symptom information directly from an informant via a semi-structured interview procedure or can be completed based on knowledge already possessed about the individual (Moss et al., 2000; Royston et al., 2025; see https://moss-pas.com).

Checklist format diagnostic tools screening that are tied to DSM or ICD psychiatric diagnoses

ADAMS

A 28-item questionnaire that screens for psychiatric disorders in people with intellectual disabilities.

Items are scored based on how they interfere with daily life (Esbensen et al., 2003).

Continued

BOX 5.2 Examples of diagnostic assessment interview tools that yield suspected DSM or ICD psychiatric diagnoses—cont'd

PAS-ADD checklist

The PAS-ADD Checklist is a screening tool designed to help identify mental health problems in people with intellectual disabilities. It can be used by family members and carers. In addition, the checklist serves as a screening tool for mental health professionals and support networks (Sturmey et al., 2005).

Why case formulations are superior to simpler diagnostic processes

Put simply, case formulation is a process by which a set of hypotheses is generated about the etiology and factors that perpetuate a patient's presenting problems, and it translates the diagnosis into specific, individualized treatment interventions

(Winters et al., 2007, p. 111)

Case formulations in psychiatric care or mental health are often made using a biopsychosocial model, in which the clinician examines all of the likely influences on the problem that prompted the evaluation (Hassiotis and Vegh, 2024). Here, we are urging the use of a developmental–biopsychosocial approach to inform assessment for people with neurodevelopmental disabilities. This includes not only determining if the patient has the symptoms listed in the DSM or ICD pertaining to a suspected mental disorder, but also understanding how the presenting problem (PP) likely came about, or the HPI (history of the presenting illness). In the case of people with ASD/IDD, the assessment should begin with an understanding of the individual's unique profile of developmentally associated strengths and challenges. Some individuals with ASD or other IDD and depression may become aggressive when asked to get out of bed in the morning, as they lack motivation to do so secondary to anhedonia. Some years ago, the observation was made that aggressive behaviors may be viewed as a "state-dependent" phenomenon in mood disorders (Lowry, 1993). Other research found that such behaviors were seen in most mental disorders, representing a sign of distress (Charlot et al., 1993). A summary of a bio-psycho-social psychiatric assessment is presented, with notes for corresponding components in a *developmental*–biopsychosocial approach.

As summarized by Shahid (2024):

> *Diagnosis is not the same as formulation! In mental health, when there is a group of consistent symptoms seen in a population, these symptoms can be categorized into a distinct entity called a diagnosis (this is what the DSM-5 does). For example, we diagnose someone with a major depressive episode if they meet 5 of the 9 symptomatic criteria. However, formulation tells us how the person became depressed as a result of their genetics, personality, psychological factors, biological factors, social circumstances (childhood adverse events and social determinants of health), and their environment.*
>
> ***(Shahid, 2024)***

Using timelines to assist in psychiatric assessment

In the process of reviewing the history of the presenting illness or problem (HPI or HPP), it can be useful to build a timeline. The timeline can include brief notes on care and other critical events, including birth and early development, special services (early intervention), schooling, medical interventions and reports, psychiatric care (including emergency services and hospitalizations), use of psychotropics over the years (hopefully with reasons for use and observed effects), as well as family events, changes in residence, relationships, any legal involvements, medical hospitalizations, and so on.

The goal is to chronologically recreate the person's journey to include occurrences that most likely contributed to their becoming the person they are today. It may also help clinicians gain a perspective on the evolution of challenges and successes. In some cases, for example, medical events in the timeline may be found next to psychiatric ones that no one ever correlated because each area was reviewed separately, as opposed to seeing them represented as they happened.

> *Case example: The individual was found to have a UTI but in a different setting than where he was treated with psychotropics for aggressive behavior. His aggressive behaviors were reduced with antipsychotic medications. However, the timeline showed he was treated for his infection at almost the same time. Over time, he remained as aggressive with antipsychotic medication use but then showed improvement later when his UTI cleared. Despite this, his care providers reported he was less aggressive due to medication. He remained on antipsychotics until a multidisciplinary consultation using a timeline demonstrated the impact of resolving his medical problem.*

Even when a reexamination of a person's challenges is being undertaken, as well as when someone is first seen for a psychiatric assessment, a detailed review of their history is warranted. These lookbacks are especially helpful when working to determine if aggressive behavior is secondary to psychiatric symptoms and disorders or to complications of one's life history in the setting of specific neurodevelopmental challenges. Timelines should include reviews of documents from the past, as often as possible written by direct observers or clinicians providing assessment and treatment to supplement patient and caregiver interviews. This might include physician notes or psychological assessments but also annual reviews, or individual education plans (IEPs), behavior plans, and behavioral data. The documented occurrences can be augmented with patient and caregiver reports.

As suggested elsewhere, an important starting point is asking the person being assessed and their caregivers when this person was last doing well. The wellness period should last for more than a day and should be sustained over a period of months or years. It is important to carefully review again what "well" looks like for this unique individual. The patient, or the parents and caregivers of youth and adults, may know when things started to become troublesome. However, they also know the positive things that support wellness. If possible, clinicians should define what this individual's *strengths are.* It is useful not to focus on what makes it easier for us to provide care but rather on what brings this person a sense of purpose or joy. What makes this person smile? What are the signs that they are happy, calm, and feeling safe and well? Then, if possible, when was this person last showing such signs for an extended period?

When the above is established and described, the patient or caregivers-informants can be asked to walk the clinician forward from wellness to the earliest signs of difficulty that eventually prompted seeking help related to aggressive and other behaviors that challenge. This may include difficulty with mood, enjoying usual activities, and biological functions. One critical tip is to insist that informants describe what they saw and heard, and not just the labels that were given to these things. Clinicians need to be open to new understandings, as sometimes past labels were wrong (e.g., a patient described as manic due to being overactive and pacing a lot without other signs and symptoms of mania, eventually found to have akathisia) (Box 5.3).

Box 5.3 Key areas in the timeline

- Family history of medical and psychiatric disorders
- Birth and Development
- Pregnancy course—term
- Birth course, term, and birth weight (BW)
- Any complications, prematurity, or low birth weight?
- Any known genetic influences or syndrome diagnoses?
- h/o genetic testing for copy number variants or anomalies including polymorphisms of CYP450 system
- Developmental course, such as when milestones were achieved.
- Age at onset of significant conditions such as a first seizure.
- Educational events/behavioral health
- Preschool era
- Early school years
- Middle School
- High School
- Early adult years
- Other Details:
  - Where was he or she along their life course to date (settings), as well as who was around, living with them, caring for or supporting them.
  - What medical, neurological, or psychiatric assessments, studies, tests, symptoms, or treatments were occurring during each time period, using dates where possible?
  - Note the onset of new medical concerns, special observations, or assessments, i.e., when the person first developed aggressive behaviors.
  - If possible, include any psychosocial or biological events that were occurring during the period.
  - Note developmental transitions such as puberty, changing schools, leaving home, etc.

Diagnostic tips for the most commonly diagnosed conditions and syndromes

In this section, a brief review of major diagnostic categories is provided. For the diagnoses most frequently found to be challenging in the Charlot et al. (2022) study about diagnostic dilemmas, one-page tips to distinguish these disorders from other causes for symptoms are provided in Tables 5.1–5.3.

Table 5.1 Anxiety disorders.

Tips for diagnosis of anxiety disorders	Background
People with ASD/IDD often display signs of anxiety in ways more closely tied to their social-emotional developmental features than their chronological age.	If a person functions from a developmental perspective in ways more commonly seen in children without a neurodevelopmental disorder, anxiety may manifest in a similar manner.
Anxiety, as seen in people functioning similarly neurodevelopmentally to a preschool-aged child with a neurotypical developmental history.	"Tantrums" or meltdowns, Fight or-flight behaviors, may be described as "irritability" when they cause or worsen aggressive behavior. Anxiety will often act as a setting event for aggressive behavior, lowering the threshold or provoking it.
Anxiety for people functioning similarly neurodevelopmentally to a school-aged child.	May be driven by a desire to connect with others but not knowing how, leading to frustration and, at times, aggressive behavior. The aggression may be a way out when there is bullying or simply exclusion from activities.
Anxiety, as seen in people functioning similarly neurodevelopmentally to a teenager or transition-aged youth.	May see an obsessive focus on having a boyfriend or girlfriend, but without understanding usual social rules and expectations, leading to frustration and anxiety, which in turn leads to aggressive behavior.
Panic is often described by informants as aggression that occurs "out of the blue."	May see what seems like a sudden onset of frantic overactivity (flight), but often with aggressive acts (fight), especially when people react severely rather than in a comforting manner.
Trauma is a very common driver of anxiety and panic.	Trauma may be associated with things not usually traumatic for people with a neurotypical developmental history, such as being teased, excluded from social groups, feeling unable to make choices, fearing transitions, and when aggressive, repeated moves occur, amplifying the anxiety.
Patients with ASD, especially, but also other IDD, often show anxiety when things change without warning or when there are transitions.	People with ASD/IDD may especially need to know what is next due to ongoing anxiety that something bad may happen. This can be described as paranoia, though it is really fear based on past experiences where bad things have happened often, and the person with ASD/IDD feels they do not have control.
There may be anxiety-driven aggression when caregivers are preventing a person from trying to leave an area, causing them severe anxiety.	Caregivers will focus on the behavior as "escape-motivated" and miss the need to determine what the person is experiencing as aversive and why, as addressing this can avert the fight-or-flight response and the attendant aggressive behavior. Example: A student with ASD cannot cope with the lesson and needs help, does not know how to ask, and becomes increasingly anxious. The intervention is to provide support or alter the task.

Table 5.2 Bipolar disorders.

Tips for diagnosis of bipolar disorder - Mania & other features	Background
For an accurate diagnosis, the person must have at least some **expansive mood.** Bipolar disorder (BD) is unlikely if a person only shows irritability (which is *nonspecific* for Bipolar and seen in many other conditions).	Irritability is common in many conditions—it is a poor diagnostic indicator without significant supporting data for the syndrome (as outlined below). Mood dysregulation is a common consequence of executive function deficits, which are common in ASD/IDD. People may be labeled with Bipolar Disorder due to periods of becoming agitated off and on.
Must observe a **high level of energy**, NOT simply poor sleep at night while being tired or napping during the day. Informants often only report night sleep unless specifically asked whether the person appears tired or naps during the day.	During mania, truly affected individuals most often have limited sleep but **high levels of energy**, appearing driven and elevated even without sleep or after minimal sleep. The overactivity tends to be goal-directed rather than random, as with encephalopathy.
Must-see **EPISODES** (minimum 4 days but usually much more) and not a chronic course (the person has been like this for years with some periods of heightened symptoms). Informants will report "cycles" when simply the person experiences some "ups and downs" in challenging behaviors and negative mood, without other BD symptoms cooccurring.	Many people with IDD are labeled with Bipolar Disorder when they don't sleep well, are restless and aggressive, but they are tired and nap during the day. The Bipolar label may derive from seeing episodes as motor agitation, which can be caused by many varied problems.
Must see *inter-episode recovery* to at least something approaching baseline—some sense of improvement. Informants may think of BD despite the person having had the same problems since age 2, often in the setting of an ASD. Chronic issues starting at a very young age usually are not due to BD.	Many people with IDD are described as "cycling," but they simply have periods when they are more stressed or display more aggression and do not have clear manic or depressive episodes. Influences on periods of increased distress are sometimes underappreciated.

Continued

Table 5.2 Bipolar disorders—cont'd

Tips for diagnosis of bipolar disorder - Mania & other features	Background
Look carefully for other explanations if the **age at onset** was during early childhood.	The US diagnoses pediatric bipolar disorder many times more than other countries like the UK due to being overly loose with diagnostic criteria for syndrome features. Many symptoms are *sensitive* but not *specific.* When symptoms have occurred since early childhood, it is unlikely to be Bipolar Disorder.
Consider other, more common reasons for aggression and alterations in mood. *Externalizing behavior/Aggressive behaviors are not diagnostically specific*	Aggression and other externalizing behaviors are a final common pathway for distress in patients with few ways to demonstrate this. Much more common causes are variable and differ from one situation to another (see functional assessment of the context of challenging behaviors). If a person has intermittent periods of increased frustration, anxiety, irritability, and aggression, this may be misinterpreted as an episode of mania. Another cause is akathisia or anxiety.
Note that **"hypersexual" behavior at times is actually sexual frustration** for young adults who have no access to appropriate opportunities.	People with IDD are often seen as needing to be asexual, even when they are adults and clearly want relationships. Because some people lack social awareness, manifestations may seem "inappropriate" and are then judged as psychopathological. In some cases, people with ID recapitulate instances of historically being sexually abused. When encephalopathic, this can also be a sign of disinhibition secondary to reduced frontal activity.
It is very rare to have manic episodes only; you should see if there were also depressions, and whether the **depression** is not better explained as a health condition.	Most people with bipolar disorder have a history of episodes of depression. It should raise suspicion if there are only reports of "mania" and never any signs of depressive episodes.

Table 5.3 Psychosis.

Tips for diagnosis of psychosis	Background
Individuals who have no way of reliably conveying what their thoughts are may not be diagnosed with a *thought disorder.*	Be skeptical of labels of psychosis in people with limited functional communication skills, regardless of their ID diagnosis or whether they speak or not.
People with approximately age-equivalent developmental, cognitive, and social cognition abilities of age 7 or less.	The critical basics of diagnosing disorders of thought include the capacity to describe one's thoughts. Caregivers often do not understand this.
Key to accuracy is identifying certain core features and not relying on others—ones that are harder to identify accurately. *Difficult: Bizarre behavior* *Better: Clear evidence of auditory hallucinations, well described.*	Research shows that language skills correlate with the identification of psychotic symptoms. Negative symptoms of schizophrenia are also easily seen as baseline features in many people with specific profiles and neurodevelopmental conditions.
Delusions are often misdiagnosed in patients with IDD, who lack an understanding of real things versus things wished for, fantasized, or *fears of* what could happen compared to what *has* happened.	Difficult to know if belief is a misunderstanding—differences between reality and fantasy (common even to some people with mild ID). Do not confuse "wishes" with delusions, i.e., wishing the person at the store counter wants to marry me because she was nice. Cognitive challenges may cause people to have loose associations due to a basic lack of appreciation for logical relations between topics—trauma effects also cause issues.

Continued

Table 5.3 Psychosis—cont'd

Tips for diagnosis of psychosis	Background
Hallucinations are often = misdiagnosed as manifestations of developmentally driven phenomena, such as: -Being socially unaware that certain statements sound odd. -Processing information out loud -Reenacting conversations out loud -Referring to people whom listeners don't know, not recognizing this -Not knowing the difference between one's own internal voice when thinking and "hearing voices." -Wanting to please interviewers—doctors—by saying yes to questions like, "Do you hear voices?" -Concrete thinking asked if you hear voices, you do. -Learning that saying "voices told me to do things" versus saying "I was angry" -Learning that saying "I hear voices" gets one a lot of concern -Not understanding reality-fantasy distinctions	If reported as only visual, these are usually not schizophrenic. Be careful; these are not secondary to a lack of appreciation of differences between one's thoughts, recollections, and actual "voices" in one's head. The question "Do you hear voices?" may even be too abstract for some individuals, who simply learn to or tend to answer *yes*. Concrete thinking may cause misinterpretations, even saying, "Yes, I hear voices." When asked, "Whose voice do you hear?" the answer might be, "Yours." Wanting to please the examiner may lead to reporting "Yes" when asked if you hear voices. Many people with IDD lack social awareness and speak out loud to reenact prior discussions or express things that are worrying them. Trauma-related anxiety, associated thoughts and worries, and nightmares often end up being seen as schizophrenic or delusional. Adults with IDD may present similar diagnostic confusion regarding the accurate identification of hallucinations when a person has cognitive challenges similar to those of young neurotypically developing children.
Ideas of reference	People, sometimes even with mild ID, will become confused about whether or not they are part of things they wish they were part of, i.e., one woman we saw pretended to be part of *Murder, She Wrote and was* seen as psychotic. She was simply lonely and bored. Help informants apply a developmental understanding. Note, too, that typically, people with psychotic symptoms in the context of schizophrenia are generally tormented and quite dysphoric.

Table 5.3 Psychosis—cont'd

Tips for diagnosis of psychosis	Background
Paranoid ideas are often misdiagnosed when, in reality, the person has IDD: • Has a history of trauma and now expects cruelty • Misses family or has grief • Feels like people are talking about them, and they are, in fact, or often do	Many times, when people with IDD have been discussed, monitored, and directed by others, they may resent this and complain, i.e., "People are talking about me, against me," etc. This can be seen as evidence of paranoid ideas when, in fact, people are talking about them. When people with IDD have trauma histories, they may expect people to be mean to them, and not because they are "paranoid."
Age of onset	Early signs of agitated behaviors, later labeled psychotic, are often a sign that it is more likely a neurodevelopmental disorder, not schizophrenia.

Adjustment reaction to an acutely stressful circumstance:
Many individuals with IDD unfortunately have "vulnerabilities" or neurocognitive and social-emotional challenges as part of their baseline profile, which result in a risk for aggressive responses to stress. When individuals who have a pattern of reacting to stress behaviorally become older and stronger, their reactions may be perceived as more dangerous or may place them and others at risk. This often results in referrals for mental health assessment. At times, the person may be displaying a lifelong tendency to react to stress behaviorally but may seem more disturbed because of their weight and size. In the latter instance, if the stress is not accompanied by or driven by a psychiatric condition, then the cause of the stress needs to be addressed, as well as how the individual responds. A general key may be the clinical course and history, so a person who began responding as they are now at about age 2 probably does not have an acute psychiatric disorder that explains the presenting problem.Medical problems or medication side effects that are associated with discomfort, odd sensations, pain, or distress can provoke alterations in both mood and behavior, or cause an acute increase in the intensity of symptoms associated with an existing mental illness. These stressors will often lower the risk of aggressive behavior. Many other contextual influences can cause an adjustment reaction. A more intense version of what the person has always done when distressed may be caused by one or more of such variables. The Recent Stressors Questionnaire (copy provided at the end of Chapter 5) can be used to generate a list of possible stressful events. The RSQ was developed because people tend to underreport things that have actually been instrumental in the need for help. Items in the RSQ are provided in Box 5.4.

Trauma

A history of highly stressful life experiences and potentially traumatic events is so common in the population that it is difficult to find people with ASD/IDD referred for a behavioral health assessment who have no such histories. Chapter 7 reviews this critical area in detail. The RSQ focuses on recent stressors, but even events from the remote past have a current influence on a person's mood and behavior. In addition, "stressors" causing major upsets for some individuals with ASD/IDD may seem small and unusual compared to the things likely to provoke a major psychiatric crisis referenced in studies of people without a neurodevelopmental disability. For example, a person with ASD and an anxiety disorder who is responding to the assignment of a new van driver might become dangerously aggressive. In fact, research suggests that just having a lot of changes may be extremely stressful for many people

BOX 5.4 Recent stressors questionnaire (items included are listed here).*

Date **Name of person providing info:**

Please check YES or NO for each item, and describe any item answered YES. Add notes if helpful.

_____Please provide details, including dates if possible.
_____Which of the following has occurred in the past year?
_____Changes in residential staff:
_____Changes in school or day/vocational staff:
_____A move to a new living situation:
_____A change in day program, job, or school/classroom assignment
_____Changes in the level, rate, or type of contact with family or significant people.
_____Illness of a loved one, caretaker, friend, or peer.
_____New peers at day/school residence or the loss of peers from these settings.
_____Changes in the level, rate, or type of contact with family or significant people.
_____Illness of a loved one, caretaker, friend, or peer
_____New peers at day/school residence or the loss of peers from these settings.
_____New Behavior Support Plan
_____Changes made in the Behavior Support Plan
_____Staff may not have been following the BSP consistently.
_____There was suspected abuse
_____Vision or hearing loss or change
_____Loss of mobility or decreased mobility.
_____Changes in doctors, therapists, teachers, or other key service providers
_____Things are different at home, work, or school.
_____The individual seems as if he/she might be ill, in pain, or uncomfortable.
_____Changes in bowel or bladder habits (new incontinence or other changes in habits).
_____Weight loss or weight gain.
_____Change in appetite or the start of a new diet
_____Any new or unusual movements of any kind.
_____Other changes in routines, even small ones, that might affect this person.
_____Hospitalizations
_____Emergency Department Visits
OTHER—PLEASE LIST

*Developed by Lauren Charlot.

with ASD/IDD who are receiving *mental health* services. Having been bullied, teased, and left out of usual positive social experiences in school are common. The onset of emotional and behavioral challenges in the teen years can sometimes be directly the result of the accentuation of such experiences. Losses, changes, neglect, and physical abuse are also all too common in the histories of people with ASD/IDD and should always be considered as possible drivers of misery and aggressive behavior. Regarding the latter, many stressors simply lower the threshold for behaviors of concern and emotional meltdowns, so that something stressful or aversive that a person might tolerate when feeling well becomes intolerable when there have been too many disruptive experiences. People with ASD/IDD and trauma histories may simply be reacting to low-level threats and stressors in large ways, as is common in the case of most people who have traumatic histories. In some instances, an inpatient stay could help in the development of acute strategies for responding to dangerous behaviors in the context of acute anxiety, panic, or *fight-or-flight* reactions to small stressors. Most youth are traumatized by removal from the family home. Data from work in the START program showed that people with a history of facility-based care were more likely to require emergency psychiatric ED care. Being placed in facilities or residential schools can be associated with traumatic experiences for many children and adults (McNally et al., 2021).

Informants or the individual may report experiences that were highly stressful, caused distress or anxiety, or that the person may have been trying to emulate bullies by becoming more aggressive toward others (as a means of self-protection).

Observers or reports may reveal links between stressors and changes in mood and behavior. The most common adult psychiatric disorders evolving from a childhood trauma history are depression and anxiety. It may also have signs of autonomic overarousal or supersensitivities to various sensory experiences, such as being highly agitated by a sudden noise or a person touching them when this was not expected.

Caregivers sometimes see intense seeking of support as "attention-seeking" without focusing on why the person needs this at the time (see detailed discussion in Chapter 9). Attention may be sought as a means of finding relief from extreme feelings of anxiety and worry. Assaulting caregivers may cause them to remove the person from the situation they are in. This may be a relief for that person, in some cases, from anxiously reexperiencing trauma. This is when there is withdrawal and escape-motivated aggression. It is critical to assess and determine why a person finds certain events, people, or environments to be aversive. Fight-or-flight types of aggression or meltdowns are sometimes mislabeled as oppositional behavior or noncompliance. See Table 5.4 for a quick assessment of recent stressors that may contribute to the risk of aggressive behavior.

Caregiver burnout

is a product of a transaction between the skills, resources, and supports the caregiver has and the degree of challenges presented by the individual. Often, people focus only on what is wrong with the "patient" when much of the problem is due to other factors, such as the caregiver's approaches or a lack of needed resources. It becomes clear when the incident leading to an acute evaluation is just another in a very long series of similar challenges, and the caregiver is clearly exhausted. On review, it may be revealed that the caregiver has few or no supports, or the person for whom they provide care is inadequately served. See Chapter 7 for more details about caregiver trauma. See Chapter 8 for more about caregiver trauma.

Anxiety disorders

Anxiety can provoke "fight or flight" responses and is very often an underlying driver of agitated, aggressive, and disruptive behavior displayed by individuals with ASD or other IDD. In many of these individuals, the combination of sensory sensitivities and typical ASD-associated challenges may set the stage for highly disruptive or even dangerously aggressive behavior that is actually the surface manifestation of a panic attack. Frequently, though antipsychotic medications may tamp down or dampen the heightened arousal state associated with these meltdowns, they do not directly help the underlying mechanisms driving the distress. Focusing on the aggression at times leads to long-term ineffective treatment with antipsychotics for the "irritability." When aggression is automatically seen as a sign of *irritability*, anxiety as a source may be missed. Anxiety may drive intermittent upsets and also be mislabeled as representing mania in the setting of bipolar disorder. When anxiety is identified and becomes the treatment target, lasting positive effects may be achieved using antianxiety medications with multiple modalities of nonmedical anxiety-reducing interventions such as preventive use of relaxation, other sensory modulation tactics, modified CBT, and many other strategies (use of transition tools, providing controlled choices, building in predictable structure or routines, teaching people to accept some change as not being frightening, etc.).

Depression

Often missed when the person is also agitated, irritable, or aggressive, these symptoms are common correlates of depression and may be viewed as surface manifestations of dysphoric states (how the person expresses distress and misery).

A family history of mood disorders of any type may add risk for depression. Some prodrome or buildup occurs, and depressive episodes do not usually have an abrupt onset. Significant depressions happen to people at an older age who have no history of mental health disorders, so it is especially important to look for medical causes in elderly patients. Symptoms happen together and clearly represent a departure from what is usual for this person, causing distress/suffering and an inability to carry on as usual. These patients may appear anxious during an interview or frankly sad. The lack of ability to be happy and loss of interest are often clear during an interview. Body posture and motor behavior may demonstrate slowing, latency to respond, and also anxiety-restlessness or psychomotor agitation. In early studies, we found many patients demonstrated both slowing and motor agitation. Depression does not generally cause loss of alertness or a clouded sensorium, but differentiating these states requires caution and, as usual, knowing how the individual processed information when well. Anhedonia is the most key and core feature, easy to ascertain. Crying and sad affective states, as well as irritability, also occur. Reduced functioning from baseline due to just not caring or low energy may be reported. Being tired a lot, disrupted sleep, but with low energy, as well as appetite and weight changes, are frequently seen, with appetite and weight going up or down. When severe, the person may speak less and may stay in bed a lot, or may seem needy, clingy, and very anxious (anxious depression may be more severe). Statements may reflect a low self-image. Many depression symptoms can also be seen when the person feels ill, or an illness is causing significant distress, discomfort, and sedation. Acuteness of onset may help point to some causes, such as an acute infection (e.g., UTI) or an adverse drug event (Charlot et al., 2017).

Mania

In the mental status examination, the patient should demonstrate pressured or rapid speech (talking very fast) or rapid vocalization that is *very different from their usual behavior.* Some people with ASD/IDD have a chronic tendency to speak rapidly, while others do so when anxious. Mood and affect should reveal at least some degree of expansive mood, far out of proportion with usual signs of happiness. Though there is often an irritable mood, it reduces the likelihood of a bipolar disorder manic episode diagnosis when only irritability is observed. As suggested, many people will sleep less but still have high energy. Caregivers may only report night sleep and fail to mention when the patient clearly appears tired and is often napping during the day. Caregivers will report being tired themselves from trying to support a person during an episode (Charlot et al., 2022; Kaplan, 2012) (see Table 5.2).

Psychotic disorder

This is not associated with a mood or medical disorder, is infrequent, and often overdiagnosed. It is important to understand an individual's logical thought capacity at baseline to avoid misinterpreting developmentally driven atypical speech or ideas as psychotic. Processing thoughts and traumas out loud without recognizing that it seems odd to others, or automatically saying "yes" when asked "do you hear voices," may be mistaken for psychosis (Dosseter, 2007). Wishing for things and talking as if they were true may not be delusional but may instead reflect poor self-image and reduced hope for a typical life—for example, claiming one has a wife when this isn't true but is deeply wished for, or a traumatized youth saying they are in a gang because they perceive these individuals as strong. See Table 5.3.

Delirium

This is also referred to as ***altered mental status*, *acute confusional state*, *toxic metabolic state, or encephalopathy.***

Delirium may be more common than most people realize when individuals in the population we support are receiving multiple medications (Lipowski, 2004). Acute onset of global changes in mental status may be a sign of a delirium-like condition. Many varied causes include adverse drug effects, acute medication discontinuation syndromes, infections, dehydration, and liver or kidney failure. Certain indications may suggest neuroleptic malignant syndrome (NMS) or NMS spectrum disorder, or serotonin syndrome. For some individuals with IDD, there may be a need to differentiate it from certain epileptic states via EEG. An affected person may appear very ill, with abnormal vital signs that can suggest the cause of the delirium (e.g., hypotension, infection).

The following are features that may be seen.

- *Dysautonomia*
- Acute movement disorder symptoms
- Odd acts like "lint picking"
- Perseverations that are generally illogical, unlike manic overexcited activity
- Waxing and waning sensorium (person is "in and out of it")
- Often restless or agitated demeanor with nonpurposeful motor activities
- Confusion that is atypical for the individual (a departure from baseline when well)
- Disturbances in the sleep-wake cycle

People with IDD, elderly people, and people with multiple medical problems are at the greatest risk. Delirium in high-risk patients can be provoked by "seemingly minor precipitants." The initial presentation of psychotic disorders, as well as dementia, can be mistaken for delirium (and vice versa). Distinctions are extremely important, with many delirium states being reversible. Note, however, that mental status changes may take a long time to recover, as one might see in the elderly; i.e., UTI-related confusional state lingers past effective antibiotic treatment for the infection, or valproic acid-related encephalopathy-provoked mental status changes linger past discontinuation of the drug (Javaid, 2024).

Dementia

Dementias tend to evolve in a slow, progressive manner and develop over months to years. It is uncommon under age 50 in typical population studies but may have an earlier onset in syndromes associated with IDD (especially Down syndrome). Incidence increases with age. Appearance and behavior vary with the degree of dementia and specific type. The progressive nature of dementia helps differentiate it from delirium or toxic metabolic states. In dementia, key features include a tendency to retain longer-term memories but struggle to consolidate new learning, with deterioration of working memory, processing speed, and other executive functions over time, as well as a gradual loss of basic skills. Some causes of dementia are reversible, but many are not.

Contributions from other acute, reversible medical problems must be ruled out based on history, examination, and various studies, including imaging and laboratory testing (Deb et al., 2022).

References

Allen, D., 2008. The relationship between challenging behaviour and mental ill-health in people with intellectual disabilities: a review of current theories and evidence. J. Intellect. Disabil. 12 (4), 267–294.

Beasley, J.B., Barnhill, L.J., Caoili, A. Eds., 2021. Integrated Mental Health Treatment Guidelines for Prescribers in Intellectual and Developmental Disabilities, 1st. National Center for START Services, Institute on Disability, University of New Hampshire, https://iod.unh.edu/sites/default/files/media/2022-05/integrated_mental_health_treatment_guidelines_for_prescribers_in_intellectual_and_developmental_disabilities_2021_0.pdf.

Beasley, J.B., Kalb, L., Klein, A., 2018. Improving mental health outcomes for individuals with intellectual disability through the Iowa START (I-START) program. J. Ment. Health Res. Intellect. Disabil. 11 (4), 287–300.

Beasley, J.B., Klein, A., Weigle, K., 2016. Diagnostic, treatment and service considerations to address challenging behavior: a model program for integrated service delivery. In: Rubin, I.L., Merrick, J., Greydanus, D.E., Patel, D.R. (Eds.), Health Care for People with Intellectual and Developmental Disabilities across the Lifespan. Springer, pp. 1629–1644.

Branford, D., Gerrard, D., Saleem, N., Shaw, C., Webster, A., 2019. Stopping over-medication of people with intellectual disability, autism or both (STOMP) in England part 1–history and background of STOMP. Adv. Ment. Health Intellect. Disabil. 13 (1), 31–40.

Buie, T., Campbell, D.B., Fuchs III, G.J., Furuta, G.T., Levy, J., VandeWater, J., … Winter, H., 2010. Evaluation, diagnosis, and treatment of gastrointestinal disorders in individuals with ASDs: a consensus report. Pediatrics 125 (Supplement_1), S1–S18.

Charlot, L., Abend, S., Ravin, P., Mastis, K., Hunt, A., Deutsch, C., 2011. Non-psychiatric health problems among psychiatric inpatients with Intellectual Disabilities. J. Intellect. Disabil. Res. 55, 199–209.

Charlot, L., Benson, B.A., Tasse, M.J., Fox, S., Hassiotis, A., 2017. Depressive Disorders. In: Fletcher, R.J., Barnhill, J. (Eds.), Diagnostic Manual – Intellectual Disability-II (DM-ID 2): A Textbook of Diagnosis of Mental Disorders in Persons with Intellectual Disability. NADD Press, pp. 165–186.

Charlot, L., Deutsch, C., Alberts, A., Hunt, A., Connors, D., McIlvane, W., 2008. Mood and anxiety symptoms in psychiatric inpatients with autism spectrum disorder and depression. J. Ment. Health Res. Intellect. Disabil. 1, 238–253.

Charlot, L., Deutsch, C.K., Fletcher, K., McIlvane, W.J., 2007. Validation of the mood and anxiety semi-structured (MASS) Interview for patients with intellectual disabilities. J. Intellect. Disabil. Res. 51, 821–834.

Charlot, L.R., Doerfler, L.A., McLaren, J.L., 2020. Psychotropic medications use and side effects of individuals with intellectual and developmental disabilities. J. Intellect. Disabil. Res. 64 (11), 852–863. doi:10.1111/jir.12777.

Charlot, L.R., Doucette, A.C., Mezzacappa, E., 1993. Affective symptoms of institutionalized adults with mental retardation. Am. J. Ment. Retard. 98 (3), 408–416.

Charlot, L.R., Hodge, S.M., Holland, A.L., Frazier, J.A., 2022. Psychiatric diagnostic dilemmas among people with intellectual and developmental disabilities. J. Intellect. Disabil. Res. 66 (10), 805–816.

Charlot, L., Sheehan, R., Hassiotis, A., 2019. Use of medications in the treatment of aggressive behavior. In: Matson, J.L., (Ed.), Handbook of Intellectual Disabilities: Integrating Theory, Research, and Practice. Springer, pp. 661–681.

Cooper, S.A., Smiley, E., Morrison, J., Williamson, A., Allan, L., 2007. Mental ill-health in adults with intellectual disabilities: prevalence and associated factors. Br. J. Psychiatry 190 (1), 27–35.

Crocker, A.G., Prokić, A., Morin, D., Reyes, A., 2014. Intellectual disability and co-occurring mental health and physical disorders in aggressive behaviour. J. Intellect. Disabil. Res. 58 (11), 1032–1044.

Deb, S., Strydom, A., Hithersay, R., Gomiero, T., De Vreese, L.P., Janicki, M.P., Jokinen, N.S., Service, K.P., 2022. Dementia in people with intellectual disabilities. In: Textbook of Psychiatry for Intellectual Disability and Autism Spectrum Disorder. Springer. pp. 719–756.

Dossetor, D.R., 2007. 'All That Glitters Is Not Gold': misdiagnosis of psychosis in pervasive developmental disorders—a case series. Clin. Child Psychol. Psychiatry 12 (4), 537–548.

Edition, F., 2013. Diagnostic and statistical manual of mental disorders. Am. Psychiatr. Assoc. 21 (21), 591–643.

Esbensen, A. J., Rojahn, J., Aman, M. G., & Ruedrich, S., (2003). Reliability and validity of an assessment instrument for anxiety, depression, and mood among individuals with mental retardation. J. Autism Dev. Disord. *33*, 617–629

Esler, A., Hewitt, A., Hall-Lande, J., Pettingell, S.L., Houseworth, J., 2019. Psychotropic medication use for adults with autism spectrum disorder who receive services and supports through adult developmental disability services in the United States. J. Autism Dev. Disord. 49 (6), 2291–2303.

Hassiotis, A., Rudra, S., 2022. Behaviours that challenge in adults with intellectual disability: overview of assessment and management. BJPsych Adv. 28 (6), 393–400.

Hassiotis, A., Vegh, B., 2024. New Insights on Challenging Behavior: assessment and Treatment Across the Lifespan. In: Matson, J.L. (Ed.), Handbook of Psychopathology in Intellectual Disability: Research, Practice, and Policy, second ed. Springer, pp. 159–185.

Hunter, R.H., Wilkniss, S., Gardner, W.I., Silverstein, S.M., 2008. The multimodal functional model–advancing case formulation beyond the" diagnose and treat" paradigm: improving outcomes and reducing aggression and the use of control procedures in psychiatric care. Psychol. Serv. 5 (1), 11.

Javaid, A., 2024. Delirium or mood disorder: the role of relative in correcting the misdiagnosis in an intellectual disability patient. PNP Int. J. Psychiatr. Nurs. Res. Advance online publication.

Jeglum, S.R., Cicero, A., DeBrine, J., Livingston, C.P., 2024. Emergency department utilization due to challenging behavior in children and adolescents diagnosed with autism spectrum disorder. Behav. Sci. 14 (8), 669.

Kaplan, S.L. (2012). "The Misdiagnosis of Bipolar Disorder in Children with Autism and Other Developmental Disabilities." *Psychology Today*.

Leyfer, O.T., Folstein, S.E., Bacalman, S., Davis, N.O., Dinh, E., Morgan, J., … Lainhart, J.E., 2006. Comorbid psychiatric disorders in children with autism: interview development and rates of disorders. J. Autism Dev. Disord. 36 (7), 849–861.

Lindgren, S., Lauer, E., Momany, E., Cope, T., Royer, J., Cogan, L., McDermott, S., Armour, B.S., 2021. Disability, hospital care, and cost: utilization of emergency and inpatient care by a cohort of children with intellectual and developmental disabilities. J. Pediatr. 229, 259–266. doi:10.1016/j.jpeds.2020.09.076.

Lipowski, Z.J., 2004. Delirium in learning disability: case series and literature review. J. Intellect. Disabil. Res. 48 (8), 713–720.

Lowry, M.A., 1993. A clear link between problem behaviors and mood disorders. Habilitat. Ment. Healthc. Newslett 12, 105–110.

Lunsky, Y., Balogh, R., Durbin, A., Selick, A., Volpe, T., Lin, E., 2018. The mental health of adults with developmental disabilities in Ontario: lessons from administrative health data. Healthc. Q. 21 (1), 6–9.

Lunsky, Y., Bradley, E., 2007. Dual diagnosis or dual confusion: limitations when utilizing non-specialist clinical data. J. Dev. Disabil. 13 (1), 185–190.

Magyar, C.I., Pandolfi, V., 2017. Utility of the CBCL DSM-oriented scales in assessing emotional disorders in youth with autism. Res. Autism Spectr. Disord. 37, 11–20.

Mazefsky, C.A., Oswald, D.P., Day, T.N., Eack, S.M., Minshew, N.J., Lainhart, J.E., 2012. ASD, a psychiatric disorder, or both? Psychiatric diagnoses in adolescents with high-functioning ASD. J. Clin. Child Adolesc. Psychol. 41 (4), 516–523.

McNally, P., Taggart, L., Shevlin, M., 2021. Trauma experiences of people with an intellectual disability and their implications: a scoping review. J. Appl. Res. Intellect. Disabil. 34 (4), 927–949.

Moss, S., Emerson, E., Kiernan, C., et al, 2000. Psychiatric symptoms in adults with learning disability and challenging behaviour. Br. J. Psychiatry 177, 452–456. https://moss-pas.com.

National Institute for Health and Care Excellence. (2015). *Challenging behaviour and learning disabilities: prevention and interventions for people with learning disabilities whose behaviour challenges* (NICE Guideline No. NG11). https://www.nice.org.uk/guidance/ng

O'Brien, M.J., Pauls, A.M., Cates, A.M., Larson, P.D., Zorn, A.N., 2024. Psychotropic medication use and polypharmacy among children and adolescents initiating intensive behavioral therapy for severe challenging behavior. J. Pediatr. 271, 114056.

Ritter, C., Hewitt, K., McMorris, C.A., 2021. Psychotropic polypharmacy among children and youth with autism: a systematic review. J. Child Adolesc. Psychopharmacol. 31 (4), 244–258.

Royston, R., McDermott, G., Ali, A., Cooper, V., Jahoda, A., Rapaport, P., … Hassiotis, A., 2025. Development and modelling study of an evidence-based manualised intervention: personalised treatment packages for adults with learning disabilities who display aggressive challenging behaviour in community settings (PETAL). NIHR Open Res. 5, 10.

Scheifes, A., Walraven, S., Stolker, J.J., Nijman, H.L., Tenback, D.E., Egberts, T.C., Heerdink, E.R., 2016. Movement disorders in adults with intellectual disability and behavioral problems associated with use of antipsychotics. J. Clin. Psychopharmacol. 36 (4), 308–313.

Shahid, A. Biopsychosocial Model and Case Formulation - PsychDB (2) https://www.psychdb.com/teaching/biopsychosocial-case-formulation 2024.

Sheehan, R., Hassiotis, A., 2017. Reduction or discontinuation of antipsychotics for challenging behaviour in adults with intellectual disability: a systematic review. Lancet Psychiatry 4 (3), 238–256.

PETAL Investigators, Smith, J., Baksh, R.A., Hassiotis, A., Sheehan, R., Ke, C., Wong, T.L.B., 2022. Aggressive challenging behavior in adults with intellectual disability: an electronic register-based cohort study of clinical outcome and service use. Eur. Psychiatry 65 (1), e74.

Solomon, M., Ozonoff, S., Carter, C., Caplan, R., 2008. Formal thought disorder and the autism spectrum: relationship with symptoms, executive control, and anxiety. J. Autism Dev. Disord. 38 (8), 1474–148.

Sturmey, P., Newton, J. T., Cowley, A., Bouras, N., & Holt, G., (2005). The PAS–ADD Checklist: independent replication of its psychometric properties in a community sample. Br. J. Psychiatry *186*(4), 319–323.

Tsiouris, J.A., Kim, S.Y., Brown, W.T., Cohen, I.L., 2011. Association of aggressive behaviours with psychiatric disorders, age, sex and degree of intellectual disability: a large-scale survey. J. Intellect. Disabil. Res. 55 (7), 636–649.

Winters, N.C., Hanson, G., Stoyanova, V., 2007. The case formulation in child and adolescent psychiatry. Child Adolesc. Psychiatr. Clin. 16 (1), 111–132.

CHAPTER 6

Psychotropic medications and aggressive behavior: The search for the magic pill

Lauren Charlot[a] and Angela Hassiotis[b]
[a]UMass Chan Medical School, Worcester, MA, United States
[b]University College London, London, United Kingdom

Interventions aimed at reducing or eliminating aggressive behavior in people with autism spectrum disorder (ASD) and intellectual and developmental disability (IDD) are a major focus in the care of individuals with neurodevelopmental disabilities, with medications being one of the most investigated approaches. Psychotropic medications have a history of extensive use to treat aggressive behavior in various other groups as well, including youth with antisocial behaviors, individuals with traumatic brain injury, and elderly patients with dementia (Charlot et al., 2019; Marcus et al., 2011). People with ASD/IDD are perhaps the most medicated group of all (Esbensen et al., 2009). Risk factors for aggressive behavior include developmental features or "autism-specific behaviors, cognitive abilities, and adaptive functioning" (Gohari et al., 2024, p. 2). This again suggests the need to include a developmental approach to evaluation. Other risk factors have varied from study to study but often include the cooccurrence of psychiatric disorders and residing in a facility or group residence. In a longitudinal investigation, rates of aggressive behavior displayed by people with IDD from toddlers (age 2) to adolescents (age 18) were identified. Among the 254 people followed, 54% of the youngest cohort, 69% of school-aged youth, and 42% of teens displayed aggressive behavior. About 10% of adults with IDD are reported to exhibit aggressive behavior (Royston et al., 2023).

Antipsychotics are the most frequently prescribed psychotropics to reduce aggressive and other challenging behaviors, and this is carried out off-label in a number of cases (Mandell et al., 2008; Bowring et al., 2017; Lunsky and Modi, 2018; Thapa et al., 2017; Brophy et al., 2018; Deb et al., 2025). Polypharmacy has emerged as a significant concern because it is associated with higher rates of drug side effects (Matson and Neal, 2009).

Aggressive Behavior of People with Autism and other Intellectual and Developmental Disabilities
https://doi.org/10.1016/B978-0-443-26552-5.00005-9

Psychotropic medication is the primary (or sole) intervention for many people with IDD, as noted by respected autism organizations such as Autism Speaks (Autism Speaks, Autism Speaks, n.d.).

In the study of aggressive acts among people with ASD/IDD, it is widely accepted that there are multiple pathways, making it unlikely that a single aggression intervention or "aggression medication" exists. The extensive use of psychotropic medications, especially antipsychotic medications aimed at controlling or reducing aggressive acts in people with ASD and other IDD, has become a concern in some countries more than others (Charlot et al., 2019; Hassiotis and Rudra, 2022). A 2017 multinational study noted higher antipsychotic use among youth in the United States compared to the UK, citing data from 2012 that showed antipsychotic prevalence for 15–19 year-olds was 2.53% in the United States, compared to 0.31% in the UK cohort. Organized approaches have been developed in the UK to reduce overreliance on medications to control behavior of people with ASD/IDD (Brandford et al., 2019). Less of this kind of strategy appears to be occurring in the United States. In some cases, psychotropics are used as a first-line and ongoing, long-term approach to contain or reduce aggressive behavior in people with ASD/IDD. This practice has continued even though there is no evidence base supporting their use beyond the short term (Deb et al., 2022; Hassiotis and Rudra, 2022; Hassiotis and Vegh, 2024). The tendency to rely on psychotropics, especially antipsychotic medications, may be contributing not only to ineffective treatment but also to the occurrence of more aggressive behaviors in some cases. This can happen when the side effects of medications intended to reduce aggressive behavior cause discomfort or distress for a person, who may then demonstrate their distress by engaging in aggressive and destructive acts.

Using antipsychotics to control aggression could potentially contribute to polypharmacy. This might occur when the initial use helps but then the medication becomes less effective over time. There is little information about this, as studies tend to be short term. However, there is some evidence that individuals treated are often given higher doses, and more medications are added over time (Esbensen et al., 2009). There may be effective medication augmentation strategies or adjustments that can prolong usefulness. However, in many instances, long-term efficacy without multiple added drugs or significant dose increases may not be realized. The varied causative factors that might be ameliorated are sometimes ignored when the approach is to conclude, "it's aggression," and the medical treatment is automatically medication alone. Psychotropics are unlikely to be a long-term solution to aggressive acts that are being provoked by pain, trauma, many frustrations, or, in general, when unhappiness is due to a lack of a meaningful life. For example, a young man with ASD who was not aggressive in a small program became much more aggressive when placed with several other people. He showed signs that he disliked group activities, but the program was unable to accommodate his need for more space and less forced socialization. He was treated with antipsychotics to reduce his aggressive behavior and this was minimally helpful for a short period of time. When the caretaking system is overwhelmed, this approach may be the quickest and easiest one to apply. In some situations in the United States, we found people who rarely saw their prescribing physician and received most of their pharmacologic treatment from emergency service encounters.

Most of the evidence base for antipsychotics reducing aggressive behavior was derived from short-term investigations. The first sets of studies that prompted their use for the aggressive behavior of youth with autism are known as the RUPP studies (Research Units on Pediatric Psychopharmacology Autism Network, 2002; Anon., 2005). Tapering or discontinuing antipsychotics has been found to be associated with a return of aggressive behavior (Anon., 2005). This would account, in part, for statistical differences favoring medication over placebo in studies that reported this finding. However, in many cases, the return of aggressive behavior in people switched from medication to placebo could be secondary to withdrawal effects. Although the number of weeks used for tapering antipsychotic medications has not always been statistically significant in predicting effective discontinuation, it has also been found that people with neuromotor side effects on antipsychotics experience more problems with tapering (more discontinuation difficulties) (de Kuijper et al., 2014; Sheehan and Hassiotis, 2017).

The National Institute for Health and Care Excellence, 2015 (NICE) guidance suggests that best practice includes use in the context of multidisciplinary assessment and multimodality interventions. NICE suggests that evidence supports only short-term use. As noted above, despite this, in some places, extensive overreliance on antipsychotics to manage aggressive behavior continues, at times without other forms of treatment (i.e., ABA-based) (Autism Speaks). In the United States, a helpful set of guidelines, the *Integrated Mental Health Treatment Guidelines for Prescribers in Intellectual and Developmental Disabilities,* was produced, with updates due presently (Beasley et al., 2021).

Side effects and increased risk of aggressive behavior

It remains unclear whether or not sedation contributes to reductions in challenging behaviors, including aggressive behavior as reported in the studies cited above. Sedation was noted frequently in children in investigations about aripiprazole (72% in the Marcus et al. (2011) study) and risperidone (51% rate in the Aman et al. (2002) study). Combined results from two RCTs in the RUPP series revealed that 67% of subjects experienced "somnolence." It is also noteworthy that a significant number of potential subjects who could have been included in these studies were excluded for a variety of reasons, including epilepsy. Anyone who had any antiepilepsy drug changes or a seizure at any time in the past 6 months, as well as any "significant" medical comorbidities, was screened out. During the study, participants could not take any other psychotropic medications. It seems that the youth included may not have been representative of patients diagnosed with autism who are usually referred to a psychiatrist due to aggressive behavior. In fact, a follow-up study of children in the RUPP (Research Units on Pediatric Psychopharmacology Autism Network, 2002) investigations demonstrated that the youth who had constipation or diarrhea were less responsive to the medications being studied (risperidone or methylphenidate). Both of the FDA-approved antipsychotic drugs have been approved only for short-term use in children with ASD. In a Cochrane review, the one long-term, placebo discontinuation study of aripiprazole found no differences between placebo and drug conditions in terms of relapse after discontinuation, "suggesting that re-evaluation of aripiprazole use after a period of stabilization in irritability symptoms is warranted" (Hirsch and Pringsheim, 2016). It should be noted that placebo effects are generally reported to be very high in studies, including children with ASD. Studying a sample of 184 children with ASD, despite benefits with respect to behavioral concerns, the researchers concluded that core ASD symptoms "failed to improve, and in fact subtly worsened over time" (p. 516) (Marrus et al., 2014).

A number of review papers examine the efficacy of using medications to treat challenging behaviors or aggression in individuals with IDD, with similar findings of statistically significant reductions in challenges as measured by the aberrant behavior checklist (ABC) irritability subscale (a measure primarily of aggressive behaviors that challenge). The most comprehensive reviews can be found through the

Institute of Health and Care Excellence in the UK (NICE) and in the Cochrane Library of systematic reviews and meta-analyses, each having slightly different conclusions. NICE reviews concluded that the quality of evidence supporting psychotropic medications being effective in treating behaviors that challenge is low or insufficient. In a study that examined the use of psychotropic medications in children with IDD, evidence was also viewed as low and showed only short-term benefits for risperidone. These researchers advised cautious use, following best practice guidelines in terms of first exhausting nondrug approaches (McQuire et al., 2015). Matson and Williams (2015) warned that parents are, at times, making decisions about what forms of intervention to choose for their children with ASD based on very different rationales than those generally applied by researchers, and a wide array of factors may impact this.

One study published in the Lancet reported on the clinical effectiveness of antipsychotic drugs in treating challenging behaviors in adults with IDD (Tyrer et al., 2009). The methodology used mirrored that of the CATIE studies (Caroff et al., 2011a), investigating "real-world" patients in treatment at multiple sites in Europe. The impact of haloperidol was contrasted with risperidone and placebo. Placebo was found to be as effective in the reduction of challenging behaviors as either of the antipsychotic drugs in the study.

A naturlistic study from Thailand reported that 39, or 86.7%, of children aged 2–15 treated with risperidone experienced adverse effects of the drug (Boon-Yasidhi et al., 2014).

The rates of adverse events related to antipsychotic medications in a systematic review of multiple RCTs were significantly higher for people on antipsychotics versus placebo (Deb et al., 2020). A review of 14 studies in 2015 revealed short-term efficacy greater than placebo for risperidone and aripiprazole but also noted significant side-effects, including elevated prolactin levels (SMD = 3.22, $p < 0.001$) and weight gain (SMD = 0.82, $p < 0.001$). These authors suggested that evidence for use in treating aggressive behavior overall was weak and that only short-term use was supported by available evidence (McQuire et al., 2015). In a more review, Pascucci and colleagues, in their comprehensive literature search looking at how effective antipsychotics were for treating aggressive behavior, noted, "Findings from RCTs were mixed" (p. 1). Some studies found efficacy (with olanzapine and risperidone), while others revealed no difference with placebo. However, there were many reports of concerning side effects, especially sedation, but also weight gain and metabolic side effects. Some of the studies addressed deprescribing antipsychotic medications (Pascucci et al., 2025).

Charlot and colleagues found that patients referred for specialized crisis prevention and intervention supports due to aggressive behavior and other major behavioral challenges had evidence of polypharmacy and side effects, though cause and effect could not be established (Charlot et al., 2020). Many of the reported side effects were suspected to contribute to an increased risk of requiring crisis services. During the assessment of people presenting with aggressive behavior, akathisia was difficult to identify among those referred for a specialized multidisciplinary assessment due to aggressive and other challenging behaviors (Salem et al., 2017b; Crocker et al., 2014). However, it was suspected to affect some patients who were initially described as hyperactive or manic after they displayed increased motor restlessness when antipsychotic medications were added or doses increased (Salem et al., 2017a; Deb, 2020; Charlot et al., 2022). People with IDD (including autism) who were treated with antipsychotics for aggressive behavior long term were found to have significant health conditions. These included Parkinsonian symptoms, dysphagia, urinary problems, and temperature dysregulation (dysautonomia), which were suspected to be secondary to antipsychotics (Ramerman et al., 2018).

Another worrisome aspect of the use of psychotropics to control aggressive behaviors is that, most often, decisions are made by others to continue or add medications (McLaren and Lichtenstein, 2019). In cases from the Charlot et al. (2011) study of psychiatric inpatients, on referral, caregivers described patients as noncompliant when they "refused" medication. This was done without considering the possibility that these individuals were experiencing side effects but were unable to clearly describe them. In one case, the "noncompliant" patient had urinary retention from combined benztropine and antipsychotic medication. The retention was so severe that catheterization became necessary. The patient became aggressive with the slightest provocation and in ways unlike her usual behavior when she was well. The antipsychotic was not helping to reduce her aggressive behavior, so it was slowly tapered. When benztropine was discontinued after lowering her antipsychotic dose, she gradually regained the capacity to urinate without requiring catheterization. Her mood and behavior also improved significantly.

Nonspecific signs of discomfort displayed by people with ASD or other IDD may be more likely to result in the prescription of psychotropic medications rather than simple pain relievers, such as nonsteroidal antiinflammatory drugs (Axmon et al., 2017). In many multidisciplinary assessments, the support team suggests that the first physician to consult when aggressive and other behaviors of concern are emerging or worsening is not always a psychiatrist. In many instances, a medical physician might be the first provider to see, or at the very least, medical questions should be addressed at these times (Wise et al., 2022).

Why do people take medications that are not working?

Even when some of the people involved in a person's care feel it may be helpful to reduce or eliminate an antipsychotic medication used to treat aggressive behavior, there may be resistance or complications in the process. Investigations have revealed both successes and barriers when attempts are made to taper and stop the use of antipsychotics (Sheehan et al., 2017). Factors that may cause more difficulty include staff and family fears. In addition to the worries of family and staff, a lack of multidisciplinary assessment and guidance, as well as a lack of access to nonmedical help, contribute to the lack of efforts to discontinue antipsychotics, even when the medication no longer seems helpful. Further, some people struggle with discontinuation effects, and the presence of significant movement disorder symptoms while taking the medication predicts that some people will have more difficulty withdrawing from it (Ahmed et al., 2000; Sheehan and Hassiotis, 2017). Patients seen on the specialty unit at UMass who were withdrawn from antipsychotics often struggled with discontinuation syndrome unless tapers were done slowly. In addition, many of these individuals displayed withdrawal-induced weight loss and motor restlessness. The latter was presumed to be a form of tardive akathisia. Many patients showed improved mood and a reduction in aggressive behaviors over time when the use of medication was replaced with positive behavior support and other nonmedication interventions, such as sensory modulation. These experiences were consistent with reports in the literature (de Kuijper et al., 2014; Salem et al., 2017b; Crocker et al., 2014; Hauser et al., 2022; de Kuijper et al., 2022).

In one case seen at the UMass multidisciplinary clinic, staff from a group residence feared medication reductions for the person being assessed during an event. They paged the prescriber on the weekend, reporting that "just as they had predicted," the person was much more aggressive. They were then surprised to learn that the physician had not yet put the order in place. There appears to be an impact on the lack of success in attempts to reduce antipsychotic medications aimed at reducing aggressive behavior which is related to caregivers worrying about their ability to manage a relapse or the return of these challenges. In some ways, this might be characterized as a "reverse placebo by proxy." In some cases UMass assessed, fearing a negative result seemed as strong an influence on outcomes as believing in a positive one often does. This is likely more probable when the feared outcome involves a person attacking the caregiver.

A significantly higher success rate for medication reduction and discontinuation is achieved when PBS assessment and intervention are provided as an alternative to medication (Gerrard et al., 2019). In general, tapering antipsychotics used for aggressive behavior requires that other approaches be available and used. This may include PBS, but also interventions such as sensory modulation, art and music therapy, as well as other therapies of varied types (i.e., talk therapies like Cognitive Behavioral Therapy). People may need help to improve communication skills and to learn how to ask for help and better describe sources of distress and frustration. The use of these kinds of plans is described in more detail in Chapter 9.

When starting out to reduce or eliminate antipsychotics, even for people with psychiatric diagnoses, it is important to recognize when the "symptoms" being targeted are the aggressive and other disruptive behaviors. If the treatment goal is for the patient to feel well again, rather than to do fewer negative things, interventions may be more effective over time. The ultimate goal is for the individual to return to the level of their best capacities to enjoy a quality of life. This needs to be based on an understanding of the person's neurodevelopmental profile and knowing their "story." It is useful to carefully consider how this person functioned when well and what factors supported this. One must determine if the aggressive behavior is new, worse, or varied in topography from the past. It is preferred that sedation is not the key means of reducing aggressive behavior frequency or intensity, other than for an acute crisis or until other resources can be accessed.

Many people with ASD/IDD who are aggressive have sleep difficulties. To avoid polypharmacy, it is advised not to medicate people to sleep at night if they have no activity or are sleeping during the day. It is important to recognize that medication refusals are often due to these agents causing discomfort, and signs of this may not be readily observed. Therefore it is best not to automatically label this as "noncompliance." Always ask the patient and consider the usual reasons why people do not want to continue taking a medication, especially those who may be more able to self-report. Respecting such rationale on the patient's behalf represents a significant person-centered approach.

Helping caregivers recognize when their actions, programs, and plans are what caused improvements can be helpful. In many cases, caregivers point to medications as the intervention that "worked," stating things like, "He was good on the Depakote." At times, this is followed by a thorough assessment revealing that the person was taking the medication for years, all the while presenting with aggressive behavior. When the person showed improvement and reduced aggressive behavior, it was after nonmedical intervention changes and additions were made. Somehow, these caregivers will often "thank the medication," even though the evidence points to their efforts as critical to the positive outcome. It can empower caregivers to recognize that there are ways to help people be less aggressive and distressed that do not involve medication and over which they have control.

Another helpful approach is to refrain from continuing medications with virtually no impact on mood concerns, even when the medication is known as one that will help with mood dysregulation. Mood dysregulation is often associated with aggressive behavior in people with ASD/IDD, who tend to manifest distress or moodiness behaviorally. Again, it is useful to avoid polypharmacy and only continue the medication if data support that the medication works to stabilize mood and reduce aggressive acts. It is also important to teach caregivers to understand that any drug tapered or stopped need not be replaced with another medication. As suggested here, many nonpharmacologic interventions are often yet to be tried and might obviate the need for multiple psychotropics, a practice that can increase the risk of adverse drug events and elevate rates or intensity of aggressive behavior.

After using medications to control behavior for more than a year, when it is determined they are ineffective, or if aggressive acts or other significant behaviors of concern increase, prior to adding more medications, it is important to follow recommendations made for conducting assessments. NICE and the *Integrated Mental Health Treatment Guidelines for Prescribers in Intellectual and Developmental Disabilities* (Beasley et al., 2021) might also be employed in the process. Caregivers who must face and manage aggressive acts require training and support to implement any plans that are developed, both to prevent and to safely respond to dangerous behavior.

Summary: The search for a magic bullet to treat aggressive behavior

It seems clear that aggressive behavior itself lacks diagnostic specificity when exhibited by people with limited means to express frustration, discomfort, pain, or simply a desire to control choices others are making on their behalf. A comprehensive approach seeking to explain aggressive acts should consider multiple potential influences each time there is a "flare-up" or significant change in the frequency or intensity of aggressive behavior. It would not be sufficient to have previously found one or even two factors having an impact and then to assume that these factors caused the present problem. Agitation and aggression can easily be provoked by a person with ASD being upset by a sudden change in routines on one day, and later provoked by feeling stiffness in their neck. For some people, it is just "what I do when I am unhappy." Drug side effects may cause discomfort and other problems that will drive up the occurrence or intensity of aggressive and other behaviors that challenge, or be the sole cause for them in some cases. More effective interventions may be derived from looking past the aggressive acts to determine why the individual is upset or distressed at the time and, as suggested here, not to quickly use medication as the long-term solution. In some instances, a medication side effect may increase the probability that small frustrations or triggers will provoke aggressive behavior, or it may be directly causing the person to become aggressive. Table 6.1 lists some of the more common adverse drug events (ADEs) or side effects seen in patients with ASD and other IDD referred for psychiatric care who present with aggressive behavior and other very

Table 6.1 Some example psychotropic side effects and impacts.

Effect	Associated impacts
Sedation—general effects	Resistance to cooperation with routines is often labeled as "noncompliance"
Sedation-related day sleep, sleep pattern disruption	Seen as refusing to get up and follow prompts May be medicated to achieve night sleep, even when day sleep occurs
Disrupted sleep architecture—nonrestful sleep	General signs of fatigue and lack of engagement in formerly preferred activities that require energy

Table 6.1 Some example psychotropic side effects and impacts—cont'd

Effect	Associated impacts
Nocturnal enuresis due to excessive sedation at night	Treated behaviorally often, or with more medications when caused by being so deeply asleep, urges are missed
Incontinence caused by neuromotor slowing, often on top of a congenital history of hypotonia, neurogenic bladder, etc.	Also treated as purposeful "behavior," with medical visits not immediately finding a medical cause, and often still treated with medications that aggravate the underlying conditions
Aggravation of GERD from lying down after eating.	Feeling tired and lying down frequently right after eating
Increased food drive	Seen as a "behavior" with restrictions, provoking intense frustration and aggression
Constipation, including with loose stools or megacolon	Also often seen as purposeful "holding" or focus is on loose stools, missing high bowel impactions
Urinary retention	May be caused by anticholinergic drugs given for the side effects of other drugs or as needed drugs. Retention may lead to UTIs and incontinence
Seizures—lowered seizure threshold	Many psychotropics elevate the risk of seizures in persons with epilepsy or those at risk for seizures
Weight gain	Related to increased appetite, slowing and decreased physical activity, and risk for metabolic syndrome
Motor slowing	General slowing may disrupt the ability to follow routines and lead to labels of noncompliance
Mental confusion—slowed thinking process	Compromised capacities for information processing and planning are further diminished, often highlighting the role of the medication in the occurrence
Metabolic Encephalopathy	Decreased surveillance for this and interpretation as dementia or psychosis
Drooling	Uncomfortable and embarrassing, but described as "tolerated" by those not experiencing it

Continued

Table 6.1 Some example psychotropic side effects and impacts—cont'd

Effect	Associated impacts
Tremors	Maybe confused with seizures, viewed as likely congenital even if not, or simply seen as not concerning—yet it may be lowering the person's threshold for agitation
Stiffness in arms or legs	May be mild to significant gait changes or fall risk, discomfort, and fatigue with physical exertion.
Stiffness in the neck or truncal or axial dystonia	Often missed, with side effects focusing on oral buccal dyskinesias or stiffness in the arms
Oculogyric crisis	Cases are at times described as having a seizure or simply "staring at the ceiling a lot"
Tardive dyskinesia	More often detected only if significant
Dysphagia	May not be understood as caused or worsened by drugs that impact motor systems and have Parkinsonian side effects

disruptive challenging behaviors. Identifying these and treating the person to alleviate the discomfort and distress from them is an often-missed yet critical way to reduce aggressive behaviors.

References

Ahmed, Z., Fraser, W., Kerr, M.P., Kiernan, C., Emerson, E., Robertson, J., Felce, D., Allen, D., Baxter, H., Thomas, J., 2000. Reducing antipsychotic medication in people with a learning disability. Br. J. Psychiatry. 0007-1250. 176, 42–46. doi:10.1192/bjp.176.1.42.

Aman, M.G., De Smedt, G., Derivan, A., Lyons, B., Findling, R.L., 2002. Risperidone disruptive behavior study group, double-blind, placebo-controlled study of risperidone for the treatment of disruptive behaviors in children with subaverage intelligence. Am. J. Psychiatry. 0002-953X. 159 (8), 1337–1346. doi:10.1176/appi.ajp.159.8.1337.

Anon., 2005. Risperidone treatment of autistic disorder: longer-term benefits and blinded discontinuation after 6 months. Am. J. Psychiatry. 0002-953X. 162 (7), 1361–1369. doi:10.1176/appi.ajp.162.7.1361.

Autism Speaks, n.d. https://www.autismspeaks.org/science-news/many-children-autism-medication-without-behavioral-therapy.

Axmon, A., Karlsson, B., Ahlström, G., Sandberg, J.B., 2017. Use of psychotropic medications among older people with intellectual disability in comparison with the general population: a register study. BMC Psychiatry 17, 355. doi:10.1186/s12888-017-1509-6.

Beasley, J.B., Barnhill, L.J., Caoili, A. Eds., 2021. Integrated Mental Health Treatment Guidelines for Prescribers in Intellectual and Developmental Disabilities, 1st. National Center for START Services, Institute on Disability, University of New Hampshire, https://iod.unh.edu/sites/default/files/media/2022-05/integrated_mental_health_treatment_guidelines_for_prescribers_in_intellectual_and_developmental_disabilities_2021_0.pdf.

Bowring, D.L., Totsika, V., Hastings, R.P., Toogood, S., Griffith, G.M., 2017. Prevalence of psychiatric disorders in adults with autism spectrum disorder: a systematic review and meta-analysis. Autism Res. 10 (12), 1990–2004. doi:10.1002/aur.1842.

Boon-Yasidhi, V., Jearnarongrit, P., Tulayapichitchock, P., Tarugsa, J., 2014. Adverse effects of risperidone in children with autism spectrum disorders in a naturalistic clinical setting at siriraj hospital, Thailand. Psych. J. 136158. doi:10.1155/2014/136158.

Branford, D., Gerrard, D., Saleem, N., Shaw, C., Webster, A., 2019. Stopping over-medication of people with intellectual disability, autism or both (STOMP) in England part 1–history and background of STOMP. Adv. Ment. Health Intellect. Disabil. 13 (1), 31–40.

Brophy, S., Shaw, R., Price, A., 2018. Health services use and outcomes for people with intellectual disability and mental health problems: a systematic review. J. Appl. Res. Intellect. Disabil. 31 (5), 767–781. doi:10.1111/jar.12414.

Caroff, S.N., Campbell, E.C., Sullivan, K.A., Mann, S.C., 2011a. Movement disorders associated with antipsychotic drugs: implications for the treatment of patients with intellectual disability. J. Clin. Psychopharmacol. 31 (1), 1–12. doi:10.1097/JCP.0b013e318204c5b3.

Charlot, L., Abend, S., Ravin, P., Mastis, K., Hunt, A., Deutsch, C., 2011. Non-psychiatric health problems among psychiatric inpatients with Intellectual Disabilities. J. Intellect. Disabil. Res. 55, 199–209.

Charlot, L.R., Doerfler, L.A., McLaren, J.L., 2020. Psychotropic medications use and side effects of individuals with intellectual and developmental disabilities. J. Intellect. Disabil. Res. 64 (11), 852–863.

Charlot, L.R., Hodge, S.M., Holland, A.L., Frazier, J.A., 2022. Psychiatric diagnostic dilemmas among people with intellectual and developmental disabilities. J. Intellect. Disabil. Res. 66 (10), 805–816.

Charlot, L., Sheehan, R., Hassiotis, A., 2019. Use of medications in the treatment of aggressive behavior. In: Handbook of Intellectual Disabilities: Integrating Theory, Research, and Practice, p. 661–681

Crocker, A.G., Prokić, A., Morin, D., Reyes, A., 2014. Intellectual disability and co-occurring mental health and physical disorders in aggressive behaviour. J. Intellect. Disabil. Res. 58 (11), 1032–1044.

de Kuijper, G., de Haan, J., Deb, S., Shankar, R., 2022. Withdrawing antipsychotics for challenging behaviours in adults with intellectual disabilities: experiences and views of prescribers. Int. J. Environ. Res. Public Health 19 (24), 17095.

de Kuijper, G., Evenhuis, H., Minderaa, R.B., Hoekstra, P.J., 2014. Effects of controlled discontinuation of long-term used antipsychotics for behavioural symptoms in individuals with intellectual disability. J. Intellect. Disabil. Res. 0964-2633. 58 (1), 71–83. doi:10.1111/j.1365-2788.2012.01631.x.

Deb, S., 2020. Psychotropic medication use in people with intellectual disability and challenging behaviour. Curr. Opin. Psychiatry 33 (2), 95–102. doi:10.1097/YCO.0000000000000579.

Deb, S., Jarkovský, J., Melicharová, H., Holub, D., Limbu, B., Třešňák, P., 2025. A whole population-based cohort study of the trajectory of the prevalence and the incidence of mental illness, challenging behaviour, and psychotropic medication prescribing in adults with intellectual disabilities in the Czech Republic between 2010 and 2022. medRxiv Available from: https://www.medrxiv.org/. doi:10.1101/2025.03.09.25323607.

Deb, S., Limbu, B., Unwin, G.L., Weaver, T., 2022. Causes of and alternatives to medication for behaviours that challenge in people with intellectual disabilities: direct care providers' perspectives. Int. J. Environ. Res. Public Health. 1660-4601. 19 (16). Available from: http://www.mdpi.com/journal/ijerph. doi:10.3390/ijerph19169988.

Deb, S., Nancarrow, T., Limbu, B., Sheehan, R., Wilcock, M., Branford, D., Courtenay, K., Perera, B., Shankar, R., 2020. UK psychiatrists' experience of withdrawal of antipsychotics prescribed for challenging behaviours in adults with ID and/or autism. Br. J. Psych. Open 6, 112. doi:10.1192/bjo.2020.97.

Esbensen, A.J., Greenberg, J.S., Seltzer, M.M., Aman, M.G., 2009. A longitudinal investigation of psychotropic and non-psychotropic medication use among adolescents and adults with autism spectrum disorders. J. Autism Dev. Disord. 1573-3432. 39 (9), 1339–1349. doi:10.1007/s10803-009-0750-3.

Gerrard, D., Rhodes, J., Lee, R., Ling, J., 2019. Using positive behavioural support (PBS) for STOMP medication challenge. Adv. Ment. Health Intellect. Disabil. 13 (3/4), 102–112.

Gohari, D., Schiltz, H., Lord, C., 2024. A longitudinal study of aggression in people with autism and other neurodevelopmental disabilities. J. Autism Dev. Disord. 0162-3257. doi:10.1007/s10803-024-06559-0.

Hassiotis, A., Rudra, S., 2022. Behaviours that challenge in adults with intellectual disability: overview of assessment and management. BJPsych. Adv. 28 (6), 393–400. doi:10.1192/bja.2022.30.

Hassiotis, A., Vegh, B., 2024. New insights on challenging behaviour: assessment and treatment across the lifespan. In: Matson, J.L. (Ed.), Handbook of Psychopathology in Intellectual Disability: Research, Practice, and Policy, second ed. Springer, pp. 159–185.

Hauser, R.A., Meyer, J.M., Factor, S.A., Comella, C.L., Tanner, C.M., Xavier, R.M., Caroff, S.N., Lundt, L., 2022. Differentiating tardive dyskinesia: a video-based review of antipsychotic-induced movement disorders in clinical practice. CNS Spectr. 1092-8529. 27 (2), 208–217. Available from: http://journals.cambridge.org/action/displayJournal?jid=CNS. doi:10.1017/S109285292000200X.

Hirsch, L.E., Pringsheim, T., 2016. Aripiprazole for autism spectrum disorders (ASD). Cochrane Database Syst. Rev. (6), CD009043. doi:10.1002/14651858.CD009043.pub3.

Lunsky, Y., Modi, M., 2018. Predictors of psychotropic polypharmacy among outpatients with psychiatric disorders and intellectual disability. Psychiatr. Serv. 69 (2), 201–206. doi:10.1176/appi.ps.201700032.

Mandell, D.S., Morales, K.H., Xie, M., Lawer, L.J., Stahmer, A.C., Marcus, S.C., 2008. Age of diagnosis among Medicaid-enrolled children with autism, 2001–2004. Psychiatr. Serv. 59 (3), 266–272. doi:10.1176/ps.2008.59.3.266.

Marcus, R.N., Owen, R., Manos, G., Mankoski, R., Kamen, L., McQuade, R.D., Carson, W.H., Corey-Lisle, P.K., Aman, M.G., 2011. Aripiprazole in the treatment of irritability in pediatric patients (Aged 6-17 Years) with autistic disorder: results from a 52-week, open-label study. J. Child Adolesc. Psychopharmacol. 1557-8992. 21 (3), 229–236. doi:10.1089/cap.2009.0121.

Marrus, N., Hall, L., Paterson, S.J., Elison, J.T., Wolff, J.J., Swanson, M.R., Parish-Morris, J., Eggebrecht, A.T., Pruett, J.R., Hazlett, H.C., Styner, M.A., Botteron, K.N., Dager, S.R., Schultz, R.T., Piven, J., 2014. Language delay aggregates in toddler siblings of children with autism spectrum disorder. J. Neurodev. Disord. 6, 29. doi:10.1186/1866-1955-6-29.

Matson, J.L., Neal, D., 2009. Severe problem behaviors in children with autism spectrum disorders: a review of research in assessment and treatment. Res. Autism Spectr. Disord. 3 (2), 418–428. doi:10.1016/j.rasd.2008.09.001.

Matson, J.L., Williams, L.W., 2015. Differential diagnosis and comorbidity: distinguishing autism spectrum disorders from other mental health issues. Neuropsychiatry 5 (2), 233–243. doi:10.2217/npy.15.10.

McLaren, J.L., Lichtenstein, J.D., 2019. The pursuit of the magic pill: the overuse of psychotropic medications in children with intellectual and developmental disabilities in the

USA. Epidemiol. Psychiatr. Sci. 2045-7979. 28 (4), 365–368. Available from: http://journals.cambridge.org/action/displayJournal?jid=EPS. doi:10.1017/S2045796018000604.
McQuire, C., Hassiotis, A., Harrison, B., Pilling, S., 2015. Pharmacological interventions for challenging behaviour in children with intellectual disabilities: a systematic review and meta-analysis. BMC Psychiatry. 1471-244X. 15 (1). doi:10.1186/s12888-015-0688-2.
National Institute for Health and Care Excellence, 2015. Challenging behaviour andlearning disabilities: prevention and interventions for people with learning disabilities whose behaviour challenges. (NICE Guideline 11). Available at: https://www.nice.org.uk/guidance/ng11/evidence.
Pascucci, A., Gerber, F., Besson, M., Kosel, M., 2025. Antipsychotic Medication for Behaviours that Challenge in Individuals with Intellectual Disabilities: a Clinically Informed Review. Front. Psychiatry 16, 1609408.
Ramerman, L., Hoekstra, P.J., de Kuijper, G., 2018. Health-related quality of life in people with intellectual disability who use long-term antipsychotic drugs for challenging behaviour. Res. Dev. Disabil. 1873-3379. 75, 49–58. Available from: www.elsevier.com/locate/redevdis. doi:10.1016/j.ridd.2018.02.011.
Research Units on Pediatric Psychopharmacology Autism Network, 2002. Risperidone in children with autism and serious behavioral problems. N. Engl. J. Med. 347 (5), 314–321.
Royston, R., Naughton, S., Hassiotis, A., Jahoda, A., Ali, A., Chauhan, U., Cooper, S.-A., Kouroupa, A., Steed, L., Strydom, A., Taggart, L., Rapaport, P., Han, W., 2023. Complex interventions for aggressive challenging behaviour in adults with intellectual disability: a rapid realist review informed by multiple populations. PLOS ONE. 1932-6203. 18 (5), e0285590. doi:10.1371/journal.pone.0285590.
Salem, H., Nagpal, C., Pigott, T., Lucio Teixeira, A., 2017a. Revisiting antipsychotic-induced akathisia: current issues and prospective challenges. Curr. Neuropharmacol. 15 (5), 789–798.
Salem, H., Nagpal, C., Pigott, T., Teixeira, A.L., 2017b. Revisiting Antipsychotic-induced Akathisia: current Issues and Prospective Challenges. Curr. Neuropharmacol. 1570-159X. 15 (5). doi:10.2174/1570159X14666161208153644.
Sheehan, R., Hassiotis, A., 2017. Reduction or discontinuation of antipsychotics for challenging behaviour in adults with intellectual disability: a systematic review. Lancet Psychiatry. 2215-0366. 4 (3), 238–256. doi:10.1016/s2215-0366(16)30191-2.
Sheehan, R., Horsfall, L., Strydom, A., Osborn, D., Walters, K., Hassiotis, A., 2017. Movement side effects of antipsychotic drugs in adults with and without intellectual disability: UK population-based cohort study. BMJ Open. 2044-6055. 7 (8), e017406. doi:10.1136/bmjopen-2017-017406.
Thapa, R., Moree, B.N., Storch, E.G., 2017. Anxiety disorders in children and adolescents with autism spectrum disorder: a meta-analysis. J. Autism Dev. Disord. 47 (9), 2805–2816. doi:10.1007/s10803-017-3199-3.
Tyrer, P., Oliver-Africano, P.C., Ahmed, Z., Bouras, N., Cooray, S., Deb, S., Murphy, D., Hare, M., Meade, M., Reece, B., Rao, B., Bhaumik, S., 2009. Risperidone, haloperidol, and placebo in the treatment of aggressive challenging behaviour in patients with intellectual disability: a randomised controlled trial. Lancet 373 (9657), 57–63. doi:10.1016/S0140-6736(08)61758-1.
Wise, E., Holingue, C., Klein, A., Caoili, A., Charlot, L., Barnhill, J., Beasley, J.B., 2022. Psychiatric Presentations and Medication Use in Older Adults With Intellectual and Developmental Disabilities. Am. J. Geriatr. Psychiatry. 1545-7214. 30 (1), 65–77. Available from: http://www.sciencedirect.com/science/journal/10647481. doi:10.1016/j.jagp.2021.05.022.

CHAPTER 7

The impact of trauma on aggressive behavior

Karyn Harvey
Sevita, Boston, MA, United States
Hello, It's Me, Boston, MA, United States
NextGen, Baltimore, MD, United States
IntellectAbility, Clearwater, FL, United States

Introduction

McNally et al. (2021) conducted a study of people with intellectual disabilities (ID) and concluded that people with ID are more vulnerable to psychological trauma than people without ID. The biology of trauma and trauma exposure is well established (Burke-Harris, 2018). We now understand the function of the limbic system in activating the fight, flight, or freeze response. More specifically, the amygdala is the region of the brain that sends out danger signals throughout the brain and body, flooding the vagal nervous system with adrenaline and cortisol. This activates the fight, flight, or freeze response (Hanazawa, 2022).

People can also respond with fawning or past attachment issues such as dependence, shame, or avoidance (Fischer, 2017; Ginwright and James, 2002). This process can be triggered by actual danger or by something that is perceived by the amygdala as indicating danger based on past experience. The amygdala, however, is not always an accurate reporter or assessor. People can be triggered by perceived dangers that may not actually exist. People with a great deal of childhood trauma are more easily triggered and more likely to exhibit a trauma-based response (Van der Kolk, n.d.). Often, the aggression we see, labeled frequently as a behavioral issue, is a symptom of trauma and a triggered amygdala. One of the symptom areas of posttraumatic stress disorder (PTSD) is arousal, as described in the classification system DSM-V. Specifically, a response of hyperarousal, triggered by a perception of danger, sometimes results in acts of aggression. The following stories of people with ID, trauma, and patterns of aggression illustrate these phenomena. When supported to feel safe, connected to others, and empowered to move their own lives, people can and do heal from these trauma histories, and aggression can be significantly decreased.

Aggressive Behavior of People with Autism and other Intellectual and Developmental Disabilities
https://doi.org/10.1016/B978-0-443-26552-5.00004-7

Josh's story

Josh Parker was born into chaos, as was his mother, Diane. There are generations of trauma, poverty, substance abuse, and neglect in Josh's lineage. This intergenerational trauma persisted in Josh's life; he was born addicted to cocaine as a result of persistent in-utero exposure. Research indicates that significant cognitive difficulties often result from prenatal exposure to cocaine (Gkioka et al., 2016). Josh was no exception. He struggled with learning difficulties and impulse control issues throughout his development.

When his mother was evicted from her apartment, Josh, his mother, and his sister ended up in a homeless shelter. There, bruises were seen on both Josh and his sister, and they were placed in a foster care home. He was separated from his sister. He was 2 years old. He would scream and cry for hours. That home no longer wanted him, so he was placed in another home. He was there for 2 years when bruises were found on him by the preschool teacher. The father defended himself, saying that Josh was out of control and only responded to being severely restrained. Josh entered yet another foster care home. They had 4 other foster care children and two of their own. There was a great deal of chaos in that home, and the older boys would bully the younger children. Josh remained there for many years until his foster mother passed away from cancer. The father could no longer care for the children. At that point, Josh entered a residential facility. By then, Josh had been assessed as having an intellectual disability and diagnosed with oppositional defiant disorder. The cocaine exposure and its effects had been missed. Josh was sometimes aggressive, particularly when he was teased or bullied in some way. During such altercations, he would be restrained by staff and sometimes placed in seclusion. After such treatment, he would become more aggressive, particularly toward staff, sometimes biting or kicking them. The cycle continued. He was diagnosed by the psychiatrist working with the facility as having bipolar disorder. He was placed on medication. He had a positive relationship with two older female staff members but no other connections. Every now and then, his mother would visit him. These visits were something he looked forward to, but they were erratic.

When Josh turned 21, he was placed in a program for adults with intellectual disabilities. There were three other young men in this home. One day, Josh was waiting by the window. His mother was supposed to visit, and he was excited. No matter what he had been through, his love for her and craving for that connection had never died. The phone rang. She canceled. When the direct support professional (DSP) in the home delivered the news: "Sorry, Bud, your mom can't make it," Josh saw red. He leapt at the staff. "You're lying!" he said. He attacked the DSP. That DSP pressed charges. Josh ended up in a locked facility with an assault charge. He was deemed not competent to stand trial and transferred to an institutional setting for people with intellectual and developmental disability (IDD), with forensic involvement and aggression. Josh saw himself as a criminal and continued to deteriorate. His aggression had defined him. No one saw him as someone who had experienced trauma. He was labeled as "bipolar" and now "forensically involved" and eventually, as having a "psychopathological disorder." In reality, Josh was someone who had severe and repeated trauma through his critical developmental periods, in addition to prenatal cocaine exposure. His aggression was a symptom of PTSD. Josh needed therapy. Gillies et al. (2016) found that behavioral issues, including aggression in children and adolescents with PTSD, decreased when they were able to receive psychotherapy. No therapist was willing to work with Josh. He was stuck in the symptom of hyperarousal, trying to protect himself, keep himself safe, and survive. He was perceived as manipulative, attention-seeking, and hostile. It was the last thing that he was. He was severely traumatized and exhibiting symptoms of trauma.

Trauma changes the brain

So many people with IDD and early childhood trauma have manifested such symptoms and have been so misunderstood. We know that trauma changes the brain and that people who have experienced a great deal of childhood trauma actually have enlarged amygdalas (Van der Kolk, n.d.). Therefore they are likely to be more sensitive and easily triggered. The working memory, the hippocampus, actually shrinks when the amygdala is enlarged, and thus, learning becomes difficult since working memory is key to acquiring new skills.

The critical question

The critical question becomes: what can we do to help? What can we do to facilitate healing? How can we activate that healing mechanism so that people with IDD, who have trauma histories and records of aggression, can move forward and be understood? We can train DSPs differently, we can focus on healing instead of correction, and we can teach support. It can be done with a trauma-informed support approach. It is critical for us to train DSPs about the trauma they are dealing with, as well as their own vicarious trauma resulting from working with such traumatized people. Reducing power struggles and increasing understanding and support are the goal. It is also critical to get people the therapy they need and to engage therapists in working with people with IDD, understanding that the work is both critical and doable. Then, people like Josh can get what they need to heal and move forward in their lives. They can get hope, direction, and inspiration—something we all need to heal and create a life of value. A life worth living.

What happened to Josh? How did services help Josh?

Josh was able to finally get the life he deserved. He was transferred to a community-based agency where staff were trained in trauma-informed supports. He was able to receive therapy, engage in a variety of activities, and ultimately get a job where he felt valued and needed. Josh was then able to heal. A few years later, with the social support and coaching of staff, Josh was able to find the love of his life. Lina, another person with similar issues and a past history of aggression and trauma, fell in love with Josh. He found his soulmate, and that relationship was key in facilitating his healing. Staff supported him and Lina through the ups and downs of their relationship, and both healed tremendously through their love and connection. They are now planning to move in together and have a wedding. Josh will still have bad days from time to time, grappling with memories of abandonment and occasional waves of grief, but on those days, Lina has learned to give him the extra support that he needs. And he does the same for her. They learned to read each other's cues and communicate their needs in couples counseling, arranged by their agency.

Waldinger (2024) was the fourth principal investigator in a long-term study, the Harvard Men's Study, examining all the factors that contribute to a satisfying and fulfilling life. This study spans over 75 years and continues today. They followed men from Harvard throughout their lives and men from the poorest section of Boston, conducting constant, multilevel evaluations. What they found was that the happiest and most fulfilled people were the ones who had positive relationships with a significant other. They also found that loneliness had a higher correlation with early mortality than even obesity. The saddest and most isolated people, often the first to pass away, were those who were socially isolated.

Josh was lucky, as his life progressed, to be embraced by a supportive provider agency, receive the therapy that he needed, and find someone he loved to share his life with. Relationships heal, research shows us—at least ones that are based on love and support. Unfortunately, many people with IDD are very isolated and live their entire lives never experiencing love or having a significant other. Relationships are a critical part of healing, and the opportunity to engage in one and find that right person is a human right, but one that is not always made available to those receiving services (Hingsburger, 2021).

Tyria's story

Tyria's father left her mother when she and her brother were 4 and 2, respectively. Her mother struggled in impoverished conditions to raise her two children. She sometimes worked, and she sometimes didn't. Employment was hard to come by. When Tyria was 7, her father appeared back in her life. He took her to McDonald's, then he took her and her brother out for ice cream. Then he was gone again. Several months later, he reappeared. He told them he was going to pick them up from school and take them to a fair. Tyria was excited, but it did not go the way she hoped.

Tyra was called to the pickup line with her brother when her father came. He drove and drove. Instead of driving to the county, he drove further into Baltimore City. They came to an abandoned, boarded-up building. Her father pushed them into the building.

"The fair is closed," he said. "We are going to play here and then go to McDonald's." The "playing" turned out to be hours of sexual abuse, or at least it felt like hours to Tyria. True to his word, they went to McDonald's, but Tyria was too upset to eat very much. He threatened her, saying he would hurt her if she told anyone about their "game," since it was a very important secret between them and it had to stay that way. It wasn't until over 20 years later that Tyria revealed the horrors she had been through to a therapist. This abuse went on periodically whenever her father would show back up in town. Then, on one horrific afternoon, her brother suffocated. Tyria never knew how or what happened. She just knew that her father had rushed him to the hospital, but he was gone. Her father never showed up again. This pain was unbearable for Tyria. She began exhibiting acts of aggression in school, where she was receiving special education services. She hit her teacher, hit other students, and was expelled. She was placed in a juvenile facility and was repeatedly restrained there. Her mother rarely visited her, as she did not have a car or many resources to travel to the county where the facility was located. Tyria's acts of aggression worsened. She was labeled "difficult due to challenging behaviors" and diagnosed with bipolar disorder. She was heavily medicated.

When Tyria transitioned into the adult system for individuals with IDD, she was placed in a provider agency in Baltimore County. She was labeled as "highly behaviorally intense," and she was assigned to have two staff members with her at all times due to the aggression she periodically displayed. The staff did not like her, and they made that clear to her. There were two other ladies in the home who had been there for a long time and did well in the setting. They didn't like Tyria either. Tyria would become upset, feeling abandoned by her mother, triggered by the staff, and completely alone.

Biologically, Tyria's amygdala, located in the limbic area of her brain, was frequently firing, signaling to her that she was in danger. Increased levels of cortisol and adrenaline were emitted into her brain and throughout her nervous system as a result of these danger signals. She was often in the "fight" mode due to perceiving herself to be in danger. Part of the danger stemmed from the scorn she felt from staff and her co-residents. There was nowhere she felt safe, and there was no one she could turn to. Although there was a great deal of money associated with her services, she was not receiving what she needed to heal from the severe childhood trauma she had endured. Her team had decided that she was too dangerous to receive therapy. Tyria had a behavior plan that rewarded her for positive behaviors, such as talking issues out and sharing with housemates, but she rarely earned her rewards. She was supposed to receive small amounts of money and have a monthly shopping trip, but those trips almost never happened. Her monthly trips to the psychiatrist only resulted in more and more medications as she sat and listened while staff complained about her behavior.

One day, Tyria couldn't take it anymore. A staff person reprimanded her because she ate some of the bread that was supposed to be for lunch sandwiches. Tyria grabbed the loaf and ran out of the house. It was raining. Tyria was not supposed to have any unsupervised time due to her behavioral issues. Tyria ran and ran. She ate all of the bread. Then she got cold and decided to go back into the house. She banged on the door. It was still raining. No one came to the door. She banged again. She screamed. She began cursing. She was getting wet. Finally, the staff opened the door and said: "You're not coming in. You can stay out there. You know what you did."

At that moment, Tyria wedged herself into the door, grabbed the staff's hair, and banged the staff's head into the door while cursing at her. The other staff came to the door and, after a struggle, restrained Tyria.

The police were called. The staff pressed charges, and Tyria was taken to jail. The agency discharged her and said they could no longer serve someone that dangerous. Tyria languished in jail for weeks until she was assessed as "not competent to stand trial." Then she was transferred to the facility in the state that served people with IDD charged with crimes deemed not competent to stand trial. It was a locked and highly restrictive facility. People there were frequently restrained. Tyria now felt the effects of her actions. However, instead of making her act more docile, she actually became more aggressive. The males there sometimes teased her. She fought with them. She also fought with the females. She was constantly losing privileges (which were few and far between already) because of her combative behavior.

Then Maura came along. Maura was a DSP who got it. She had been through some things herself. She was caring, consistently calm, and never acted with fear. Slowly but surely, Tyria began to trust her. She began to open up to her. They got permission to go on walks together and would walk the grounds over and over. Tyria began to lose weight and get excited about feeling healthier. Tyria began to open up. Maura convinced her to start working with a therapist. A therapist began to visit every week. Tyria still had a great deal of agitation and had repeated conflicts with certain other residents at the facility, but her overall state was significantly calmer. She was sleeping better as well.

At first, Tyria told the therapist that she would never talk about her life or what had happened to her. So instead, they started working on an autobiography, *My Book to Celebrate Me*. Tyria was so proud of her book about herself, celebrating who she was and building up a positive sense of self, that she made copies of it and shared it with all the people on her team. Maura and the therapist worked together at times in their strategy of support. Other DSPs began to follow suit. Tyria was no longer seen as the "problem person" in the program. The day came when she finally agreed with the therapist to start engaging in EMDR.

EMDR stands for eye movement desensitization and reprocessing. It was developed by Francine (Shapiro, 2012) and is a therapy designed to assist in the processing and ultimate resolution of trauma. It was the perfect therapy for Tyria because it also does not require a great deal of verbal processing or engagement. It involves bilateral stimulation and is believed to work below the conscious level. Research!

Tyria was finally able, after a year of preparatory work, to process some of the most significant traumas she had experienced, largely related to the sexual abuse and the death of her brother, through engaging in EMDR. She participated in EMDR for close to a year. Incidents of aggression continued to decrease. She was then moved to a less restrictive facility. Two years later, she was placed in a provider agency within the community. There are good days and bad days, but the bad days are fewer, and the good days have increased. Tyria is able to engage in services, live and interact in her community, and have a life where she is actively involved in daily activities and is developing genuine friendships.

Ginwright (2018) has conducted research on trauma, focusing on youth who struggle with the effects of PTSD and the ways in which they are able to heal. He demonstrated the effects of engagement, mentoring, and community involvement as key factors in healing.

Tommy's story

Tommy was born to a mother who, in all likelihood, had an intellectual disability and never received any support or services. She raised Tommy predominantly in homeless shelters after her own mother died when Tommy was 6. The shelters did not keep people for long, so Tommy frequently changed schools as he rotated shelters with his mother. When he turned 18, his mother told him that now he was on his own and she no longer had to take care of him. Tommy went to a shelter for men. The exact facts of what happened to Tommy in that shelter are not known, but there was an incident documented that ended up in Tommy's arrest and incarceration. Every morning, the residents of the shelter had to leave and could not return until 6 p.m. They could get one substantial meal at the food kitchen, but that was all they could rely on. One night, the men were lined up, sitting on the curb, waiting to enter the shelter, and someone bumped Tommy from behind. Tommy startled, freaked out, grabbed a brick that was lying in the road near him, and threw it at the person who had bumped him. In all likelihood, this was a startle response, frequently seen as a symptom within the symptom category of arousal in PTSD. The person assaulted pressed charges, and Tommy ended up in jail. There, he was assessed for the first time as having an intellectual disability. He had fallen through the cracks in the Baltimore City School System. He was deemed not competent to stand trial and was placed in an institutional setting for people with ID who had forensic involvement.

The first thing that Tommy was told when he arrived at the facility was that if he worked hard, did everything the staff told him to do, and followed all directions, he could get a placement in the community within the next year or two. Tommy immediately stopped doing anything anyone asked him to do. He would not take a shower, he would not shave or get clean clothes, he would not allow anyone to cut his hair, and he refused to shave. He had one staff member he trusted. He would talk to her but no one else. The psychologist labeled him as manipulative. He confided in the trusted staff member, Eileen, that he didn't want to go to the community. For the first time in his life, he had his own room, three square meals a day, and someone who supported him and listened to him. He didn't want to go anywhere and was doing everything he could to make sure he stayed in the safe place he had finally landed in.

The staff there received training in trauma and trauma-informed behavioral approaches. They were all determined to help Tommy feel as safe as possible, as well as to assist him in making a successful transition into a community placement when the time came, and he was ready.

They realized that he didn't feel safe taking showers. In all likelihood, negative things had happened to him in the showers at the shelters he had stayed in. Eileen came up with the idea to stand outside the bathroom door while it was shut, and he was inside, and talk loudly to him, telling him that he was safe and no one was coming. He began taking showers. They got permission to go shopping with him, and he got clothes that were not hand-me-downs, clothes that he actually chose. He kept those clothes clean and wore them proudly. He got haircuts and learned to shave with an electric razor. He eventually left the facility and moved to a community placement where he received support and engaged in activities on a daily basis. But that transition was gradual, and the staff made sure he visited frequently, met his new staff and co-residents, and decorated his own room. The process took over 6 months. He continues to reside there and is living more independently than when he first arrived. His progress is significant. His act of aggression was a trauma response. He startled and attacked. He was not the manipulative, aggressive person he was described in reports to be. He was someone with severe childhood trauma whose main concern was to be as safe as he possibly could be. Once he was understood, and his trauma was understood, he was able to receive the support he needed and move forward in his life.

Summary

As you can see from these three stories, based on actual situations, people do recover. But certain factors must be present in order for that healing to occur. One critical factor in the lives of people with IDD receiving supports is the DSP. The DSP who is able to communicate a sense of safety, particularly emotional safety and support, is the person who can facilitate healing. Relationships, as well, can facilitate healing. A DSP who can coach a person in having meaningful, unpaid relationships, be they with friends or significant others, is someone who facilitates healing. Finally, a person needs to have agency, choice, and a voice that is heard on a day-to-day basis to move forward in their lives. If others are always in control of that person's life in every aspect, it is difficult for true healing to occur. Families, DSPs, and other team members need to explore ways in which a sense of safety, connection, and empowerment can be facilitated in the lives of people who have experienced trauma. When that occurs, aggression decreases, and people are able, much like Josh, Tyria, and Tommy, to move forward in their lives with strength and hope.

References

Burke-Harris, N., 2018. The Deepest Well, Simon and Schuster.

Fischer, J., 2017. Healing the Fragmented Selves of Trauma Survivors, Tantor.

Gillies, D., Maiocchi, L., Bhandari, A.P., Taylor, F., Gray, C., O'Brien, L., 2016. Psychological therapies for children and adolescents exposed to trauma. Cochrane Database Syst. Rev. 1469-493X. 2016 (10). Available from: http://as.wiley.com/WileyCDA/Brand/id-6.html. doi:10.1002/14651858.CD012371.

Ginwright, S., 2018. The future of healing: shifting from trauma informed care to healing centered engagement.

Ginwright, S., James, T., 2002. From assets to agents of change: social justice, organizing, and youth development. New Dir. Youth Dev. Winter(96), 27–46. doi:10.1002/yd.25. PMID: 12630272.

Gkioka, E., Korou, L.M., Daskalopoulou, A., Misitzi, A., Batsidis, E., Bakoyiannis, I., Pergialiotis, V., 2016. Prenatal cocaine exposure and its impact on cognitive functions of offspring: a pathophysiological insight. Rev. Neurosci. 0334-1763. 27 (5), 523–534. Available from: http://www.reference-global.com/loi/revneuro. doi:10.1515/revneuro-2015-0064.

Hanazawa, 2022. Polyvagal theory and its clinical potential: an overview. Brain and Nerve = Shinkei Kenkyu No Shinpo 74 (8), 1011–1016. doi:10.11477/mf.1416202169.

Hingsburger (2021) https://youtu.be/yakEqrbRr0c?si=ean46wrH74051MVz.

McNally, P., Taggart, L., Shevlin, M., 2021. Trauma experiences of people with an intellectual disability and their implications: a scoping review. J. Appl. Res. Intellect. Disabil. 1468-3148. 34 (4), 927–949. Available from: http://onlinelibrary.wiley.com/journal/10.1111/(ISSN)1468-3148. doi:10.1111/jar.12872.

Shapiro, F., 2012. Getting past Your Past, Tantor.

Van der Olk, B., (n.d.) The Body Keeps the Score. Penguin Books.

Waldinger, R. 2024 https://youtu.be/8KkKuTCFvzI?si=AqMnjnT8w5SDA9Vc.

CHAPTER 8

Caring for caregivers: Supporting family and paid caregivers of individuals with ASD/IDD who display aggressive behavior

J. Leonard Peebles[a] and Lauren Charlot[b]
[a]Active Recovery Erie, LLC, Erie, PA, United States
[b]UMass Chan Medical School, Worcester, MA, United States

Introduction

Parents of people who have autism or intellectual disabilities, with or without autism, who display aggressive challenging behavior are more than just first responders—sometimes they're the only responders. First responders walk into the unknown on every call. The objective is very clear; how to get there is not. This chapter will review various techniques that can help parents provide kind, loving, caring, and calm help to their child or adult child. These strategies are also ones that paid caregivers can use.

Consider the following:

Your beautiful boy or girl, a cute, cuddly 2-year-old who is the striking image of their father or yourself, suddenly now stiffens when you hold him or her. Over an 18-month period, your toddler slowly loses motor functioning. He drops things he used to hold firmly. He no longer rolls the ball back to you. He's not interested in interacting. When he is, he bites you, slaps you, screams, and pushes you away.

Or ...

You are assigned to a family whose son has recently been diagnosed with autism. The 22-year-old boy is big and strong, now hits people, and occasionally runs into the street toward oncoming traffic.

As he does, how do you respond to this? What are your fears? What are your expectations? What skill options, or what tools in your toolbox or your tool belt, or in your pocket, do you have

Do you stand frozen?Do you start yelling and screaming for him to stop? Do you run into the road behind him? Do you run into the road in front of him toward the oncoming car? Do you grab him and spin him around? Do you hope someone

Aggressive Behavior of People with Autism and other Intellectual and Developmental Disabilities
https://doi.org/10.1016/B978-0-443-26552-5.00008-4

else will run into the traffic and stop him or spin him, or that the driver will brake suddenly or swerve into their own oncoming traffic?

My safety programs taught me … what? They didn't get that specific; they did not say, "Here's what you do." You don't take people with this history for a walk. But you were told in the treatment plan to take him for walks daily to physically provide exercise. But this is not what you had anticipated.

Do you run screaming, your hands up in the air, away from the situation, wondering why you signed up for this to begin with, or hating the people who told you what to do … or had not told you about this?

It is an impossible task to face repeated exposure to aggressive behavior. Herman (2015), van der Kolk (2014), and Dana (2018) have explained in a rather eloquent way the reactions to stressful life experiences previously thought of as mental illness or breakdown, which are now considered and understood in terms of expected, understandable human behavior as a stress response. Caregivers should recognize that it is not in any way pathological. You are not mentally ill. You may be suffering, but you are surviving. It's not that you're going crazy or losing your mind. It is an unusual, extreme circumstance that requires extreme tolerance, understanding, and effort.

Overview

People with autism spectrum disorder (ASD) and other intellectual and developmental disabilities (IDD) often present with aggressive behavior when referred for psychological or psychiatric help (Hassiotis et al., 2020). This takes many forms, including assaulting others, destructive acts to property, and self-injurious behavior, frequently requiring caregivers to respond to maintain safety. The topic of how people may come to engage in aggressive acts and ideas for intervening is addressed in multiple chapters in this book. The concern of how to understand and support the stress experienced by caregivers of these individuals has received limited attention and is the topic of this chapter. Research shows that exposure to physical and verbally aggressive behavior is strongly associated with poorer well-being for caregivers, including those who are paid (Ryan et al., 2021). The transactions between children with IDD and their parents when the children display aggressive behaviors toward their parents are complex. The responses of parents to this will impact their child's future behavior, but also their own. This continues when children become adults and remain in the care of family members. Stress related to caregiving may even persist when family members and an individual with ASD/IDD are living apart.

As noted, paid caregivers are also affected by these complex transactions and can suffer trauma from these experiences as well.

In this chapter, the significance of the stress and trauma that aggressive acts aimed at caregivers confer, as well as some possible strategies to support them, are discussed (Kalawatia et al., 2025; Reyes-Martín et al., 2024). Stress, fear, burnout, trauma, depression, and anxiety can occur in both family members and paid staff providing care (Chireh et al., 2025). Marital strain, the impact on siblings, and social isolation are all stresses affecting family caregivers. There may even be job loss for family or paid staff and challenges in maintaining consistent staffing in residential programs (Cuzzocrea et al., 2016; Ryan et al., 2021; Sánchez Amate and Luque de la Rosa, 2024). One major concern is the potential development of posttraumatic stress symptoms or compassion fatigue for these individuals. There can also be feelings of guilt, helplessness, and stigma. All such effects of caring for a person with ASD/IDD who displays aggression take a toll on the caregivers, which in turn may add risk for ongoing aggressive behavior in the people for whom care is provided. In essence, this can become a cycle of stress and trauma. The bidirectional or transactional nature of the impacts of parental stress and children's aggression has been confirmed in studies of youth with ASD (Rodriguez et al., 2019).

The extent of help and support provided by the system of care will impact outcomes. Gaps exist in training to promote expertise in working with individuals with ASD/IDD who present with serious aggressive and destructive behaviors, as well as in treating individuals with psychiatric comorbidities. Some countries have national healthcare-sponsored programs. Though these countries have many concerns too, they perform better with respect to policies aimed at preventing potential problems. In the United Kingdom (UK), e.g., there are national guidelines (National Institute for Health and Care Excellence, 2015) and national policies (NHS England, 2017) including specialist service access and crisis pathways specifically designed to meet the needs of people with intellectual disabilities who present with "behavior that challenges."

Other barriers to improved outcomes include inadequate training or supervision for direct support professionals. Families and even agency programs may need some assistance in the form of respite. There may be insufficient respite services, which are especially needed for families. Crisis services for families, apart from their family member ending up in an emergency department (ED) without medical personnel understanding the relative's needs, are often unavailable (Kalb et al., 2016). Systems may inadvertently develop policies that emphasize reactive management to stop aggressive behavior. At times, these systems also tolerate a heavy reliance on polypharmacy to control aggressive behavior without understanding and prevention through careful multidisciplinary assessment and treatment (see earlier chapters addressing this topic).

How education and training can help

Training in the areas of positive behavioral supports (PBS) and trauma-informed care often teaches proactive and de-escalation skills. Programs like this have actually been found to offer significant support. Peebles examined whether the S.O.L.V.E. Program ("Strategies of Limiting Violent Episodes"), a 20-h course teaching verbal de-escalation and noninjurious physical control, had a positive impact on trainees (Peebles, 1992). Peebles implemented a structured training program that included education about the causes of aggression in people with IDD, examining principles of behavioral assessment, prevention strategies, and tactics for safe, noncoercive de-escalation skills. Changes in staff attitudes, such as confidence in coping with aggression, self-efficacy, physical self-efficacy, and attitudes toward behavior modification, were found. Trained staff demonstrated a significant increase in confidence when responding to clients who engaged in aggressive behavior as compared with nontrained peers ($p < .0001$). The results offered early empirical support for the idea that structured, non-injurious aggression-management training significantly enhances staff confidence, a foundational step toward safer care and less trauma for the providers of the care (Miller et al., 2025; Peebles, 1991; Peebles, 1992). Miller et al. examined 22 papers that looked at interventions to train staff for responding to work-related violence in caring for people with IDD but concluded that efficacy was variable. These authors called for more research addressing the management of training and safety actions to protect staff (Miller et al., 2025). In general, emphasis on understanding the functions and root causes of aggressive behavior provides more effective means of reducing their occurrence, as do communication supports and environmental adaptations that better address the person's unique individual needs (Beneytez, 2023). Details of approaches for these areas are provided in Chapter 9.

Psychological and emotional support strategies

Counseling, peer support groups, and psychoeducation are three ways caregivers can be helped in the management of stress related to caring for a person with ASD/IDD and behavior that challenges, especially aggressive behavior. Interventions to address trauma and those aimed at building resilience are also beneficial. Included among such strategies are mindfulness and supports based on cognitive behavioral therapy (CBT).

The emerging evidence has demonstrated that acceptance and commitment therapy (ACT) can reduce parenting stress for caregivers of youth with neurodevelopmental conditions (Guo et al., 2025). Mindfulness and CBT have also been shown to reduce parental stress (Li et al., 2024). Parent training studies have been conducted where researchers measured parent outcomes in the setting of a randomized controlled trial (Li et al., 2024). Parent stress/competence, when compared to "usual care," is improved with training supports (Bearss et al., 2015).

Organizational and system supports for paid caregivers

Paid caregivers also need assistance and may become "burnt out" or traumatized when dealing with the daily aggressive behaviors of individuals for whom they provide direct care. Supportive supervision is critical, and including them in treatment planning reduces risk. Debriefing can be an important way to learn from difficult experiences while trying to keep a person with ASD/IDD who has engaged in aggressive behavior safe. Agencies and employers need to recognize that there are both practical and emotional components involved in this work. They must create safe mechanisms for reporting and foster an overall nonpunitive culture. Policy-level interventions, when sensitive to the needs associated with caregiving for this population, can reduce burnout and turnover in care (Chireh et al., 2025).

Family supports

Family members, at times, will find support from each other in organizations like Autism Speaks (a US-based nonprofit advocacy organization). This can be helpful to simply have one's experiences validated. Also, practically, families need access to care coordination for their loved ones, as well as respite care. Caregiving can be exhausting and frustrating, and this is amplified when there are no breaks and when services and supports are challenging to access (Sánchez Amate and Luque de la Rosa, 2024).

Access to some form of applied behavioral analysis-based consultation and services, as well as parent training in reasonable strategies that can be used at home, are valuable tools. A consultant may develop a positive support plan for use at home. Often, it is primarily up to family members to implement such plans. Plans for use at home work best when families are integrated into the treatment planning process. Besides PBS plans and tools, access to speech and language therapy and occupational therapy services for after-school, work, or day services is also important. Having your family member be able to follow a structured schedule of meaningful and engaging activities at home can aid in reducing aggressive behavior (Coulman et al., 2021).

Since many people with ASD/IDD who present with aggressive behavior see a number of professionals, coordination of this work with good communication can reduce stress. The professionals, such as psychiatrists, psychologists, social workers, nurses, behaviorists, and others, also require training to understand the unique needs of this client population. For example, some clinicians may label developmentally driven challenges in information processing as evidence of psychosis or identify meltdowns displayed by people with autism as evidence of mania. Treating clinicians often need to rely on informant reports when people with ASD or other IDD have significant communication challenges. When there are several treating clinicians but limited communication and coordination of care, the risk of stress and aggressive behavior is increased. The importance of a comprehensive multidisciplinary assessment process has been emphasized by specialists addressing the mental health needs of people with ASD/IDD (Hassiotis et al., 2020; Charlot et al., 2022).

When supporting people under these conditions, it is vital to recognize the individual caregiver's strengths. Ensuring that people know when they are making progress themselves will add to the efficacy of their caregiving and, in turn, to their own well-being. The use of reviews of successful outcome stories and a recovery-focused approach can be effective. It is important that a culture of compassion and mutual respect between caregivers and professionals is fostered (Hastings and Taunt, 2002; Lunsky et al., 2017). Below are a variety of tools to bolster the caregivers' sense of their strengths and to promote well-being.

Interventions targeting stress and trauma

To understand options for treating trauma symptoms, it helps to consider how and why people react to crises, such as those occurring when providing care to a person with ASD or other IDD who becomes aggressive. What is the basis for anxiety, fear, freezing, or becoming agitated when facing a violent event or threat? Neurobiological studies show that the amygdala, hippocampus, and autonomic nervous system connections are the source of the more emotional aspects of stress responses, while the prefrontal cortex and its connections mediate cognitive components. Repeated traumas weaken interconnections between these two systems, so that overriding fear and anxiety using our "thinking" brain becomes more and more difficult (Teicher et al., 2016). Overall, there is good news and bad news. The bad news is that people exposed to repeated traumas experience neurobiological changes and become negatively impacted over time. The good news is that this can be changed. Trauma-informed interventions help reduce the negative impacts of repeated traumatic life events. Those interventions that rely less on cognitive processing and those that employ this are discussed as follows.

As suggested by Sapolsky (2004) and Shapiro (2018), who described "why Zebras don't get ulcers," the traumatized person will remain activated (a ramped-up ANS) and on *alert for dangers*, ready to protect themselves. Zebras don't get ulcers because they become aroused and react to threats, but they also relax once the immediate danger has passed. Humans, on the other hand, can sustain fear and anxiety, staying overly activated persistently without periods of relief over long periods of time. In some cases, anxiety or fear-driven fight-or-flight kinds of responses are triggered without the degree of danger that was formerly required. Tactics suggested (table/box, etc.) were developed to address this dilemma so that we can know or remember stressful or traumatic events without reliving them emotionally each time. Strategies shared ideally help people be prepared to meet dangers in a calmer and more effective manner.

Bottom-up interventions for caregivers supporting individuals with ASD/IDD who exhibit aggressive behavior

Bottom-up interventions target the body's automatic stress-response systems (e.g., autonomic, sensory, and motor pathways) before engaging higher cognitive processing systems. Somatic or body-based kinds of interventions may be ideal for stressed or traumatized people caring for individuals who display aggressive behavior because exposure tends to be repeated and unpredictable. When physical threats occur in this way, there is chronic stress with a significant risk of trauma-related dysregulation in the caregiver. Bottom-up practices are important as they help restore physiological safety, which will, in turn, improve emotional responses to these adverse events (Porges, 2011; van der Kolk, 2014).

Polyvagal theory, first introduced by Porges (2011), is a popular theory and model highlighting bottom-up strategies such as breath-based interventions, the use of voice prosody, rhythmic movement, and grounding techniques through sensory input. The model suggests ways to target "dorsal-vagal" sources that drive the sympathetic nervous system's fight-or-flight responses. This essentially involves addressing "safety first" before working on cognitive controls. In this model, the more cognitive aspects of intervention are said to be related to the "ventral-vagal" system. Theoretically, this system mediates social connection, which is described as providing calming support that helps sustain a person's sense of safety (Porges, 2011). Grossman (2023) refutes the above claims from a neurobiological perspective. Despite this, approaches based on this model are often used by trauma therapists as part of a number of bottom-up strategies.

"The polyvagal theory has generated significant debate. While it is necessary to acknowledge critiques of the polyvagal theory, it is also possible to recognize the value in its conceptual framework for understanding complex phenomena in trauma and the various advantages to using aspects of this approach in clinical practice" (Giroux et al., 2023).

From other perspectives tied to neurobiological research, it is also suggested that people work to promote states of calm before trying to alter thought patterns (Murphy et al., 2022). In the case of treating individuals with ASD/IDD, as well as caregivers, the interventions below can be useful. Over time, using trauma-informed treatments, the person will hopefully develop more tolerance for small stresses and not be as distressed by them as they had been by more serious, dangerous events.

Somatic and body-based therapies

Bottom-up interventions work to increase physical self-awareness of stress and reduce arousal. They involve direct approaches to our body's trauma responses, such as startle, freeze, hypervigilance, and tense muscles. Sensorimotor treatment is designed to combine posture, movement, and sensory awareness to address trauma-triggered phenomena. Examples include flinching, bracing, or dissociation. Yoga can be adapted to address trauma and is available in many communities. This form of yoga uses light or gentle movement and is aimed at repairing one's rhythms, breath, and grounding. Yoga in this way has been described as a way to improve vagal tone and, ultimately, promote a sense of safety in the body (Levine, 2010; Ogden et al., 2006; Emerson, 2015). Breathing in a way that prolongs exhalation helps to jump-start parasympathetic regulation. Related techniques include sighing deeply (sometimes called a physiologic sigh), where the person takes two short inhales and then follows these with a long exhale. Square breathing or breathing together with individuals to get the two in sync may also be useful. An advantage of breathing techniques is that they can be quickly employed without the need for any special tools (Bentley et al., 2023).

Regulation models focus on restoring what Porges called ventral vagal states of calm. Tactics for this approach include slow exhalation breathing, prosody-based vocalization (e.g., humming, soft-tone talking), the use of warm facial expressions, a gently appearing gaze, and rhythmic movement. These practices reportedly counteract sympathetic overactivation and help caregivers regain access to connection and problem-solving abilities (Dana, 2018; Porges, 2011). Rhythmic sensory inputs help to regulate the nervous system when a person is stressed and overly activated. Slow, rhythmic walking; simply engaging in rocking movements (forward and back or side to side), bilateral sound stimulation, and tapping sequences (i.e., tapping gently on each thigh) can all contribute to stress reduction (Perry and Winfrey, 2021).

Grounding and sensory regulation strategies are likely to be most helpful when aggressive crises are occurring or immediately afterward. This is typically when a caregiver is in an over-aroused state. A variety of techniques fall under this category, including countdowns like the 5-4-3-2-1 technique, creating a quick shift in one's temperature with a cold cloth or water on the wrists, deep pressure or weighted items, and standing or "rooting" exercises that can help stabilize balance, which in turn can reduce panic. It is also possible to moderate or alter sensory features of an environment, such as reducing noise, using low lighting, and adding calming scents while avoiding those that can be aversive (e.g., noise, light, smell) (Kalb et al., 2016; Rumball et al., 2020). These approaches are similar to Occupational Therapy work with people with autism in the area of sensory modulation diets.

Top-down and other approaches

Eye movement desensitization and reprocessing (EMDR)

EMDR is a *bilateral stimulation* therapy (van der Kolk, 2014; Shapiro, 2018). EMDR is sometimes seen as having both bottom-up and top-down aspects.

Shapiro (2013) first noted that when stressed, simply moving in ways as simple as walking provided some relief. She then developed the therapy known as EMDR (this is the most user-friendly version written by Shapiro). EMDR therapy has been shown to help people with post-traumatic stress disorder and trauma-related symptoms and problems. It is an evidence-based, trauma-focused psychotherapy. A goal of EMDR is to help people process their memories of past stressful life experiences, to remember without reliving them. EMDR moves us from an emotional to an intellectual understanding. The structured EMDR model has been used for people with autism and other IDD and can be applied to caregivers as well (Shapiro, 1989; Shapiro, 2018). The eight steps used in EMDR include: history-taking, preparation, assessment, desensitization, installation, body scan, closure, and reevaluation.

In EMDR, bilateral stimulation, including guided eye movements, alternating taps, or tones, is employed during the session, while the person focuses on their memory of the trauma. It is held that the process promotes healing in the brain (using natural neurological-based mechanisms). The goal is to reduce emotional distress as well as to reform prior problematic beliefs provoked by the memories of the trauma. Increasing evidence supports the use of EMDR for complex trauma, anxiety, dissociation, and other conditions, and it has been applied in therapy for a variety of patient populations. Adapted versions have been used with people with IDD (Stekkinger-de Vries et al., 2025).

Trauma-focused CBT

One important example of therapies that are based on impacting a person's thought processes to enhance the capacity for responding to difficult situations is trauma-focused CBT. This approach can be used to assist people with trauma symptoms, such as those reported at high rates among parents of children with ASD. Stewart et al. (2017) conducted a focus group with mothers, 40% of whom described stressors. These 40% of mothers reported having trauma-related symptoms, while all participants reported some forms of stress associated with their parenting of a child with ASD (Kalawatia et al., 2025; Stewart et al., 2017).

A key goal when a person undergoes trauma-focused CBT is to provide education to the caregiver, which, in turn, leads to reduced unsupported beliefs and fears. This begins by understanding how repeated exposure to aggressive behavior can produce trauma-driven challenges tied to "fight or flight" responses. The outcomes within these interactions are not "one way." Each person's manner of responding impacts the other. How a person with IDD reacts to their stress impacts how the caregiver responds, and how the caregiver responds impacts the person for whom care is provided. The unwanted result could be a cycle of negative reactions between the pair in such situations. This can evolve into a downward, debilitating spiral.

Often, caregivers may start to think that the person for whom they are providing care is intentionally being threatening or aggressive to "get them." This kind of thought pattern is unhelpful and may even increase distress for both parties. Instead, caregivers can be supported to replace such thoughts with an understanding of the true sources of difficulty, such as the function of and influences on the behavior (e.g., medical or traumatic histories). Initial sessions may employ grounding and relaxation strategies (e.g., breathing techniques that increase a sense of calm). It is most helpful when people learn to recognize their own signs of building stress and are able to act early. A very important aspect of this form of CBT is *reframing*. The traumatized person needs to reframe the meaning of crisis events to prevent feelings of shame or guilt and to avoid "catastrophizing." Safety planning using strategies such as those outlined in PBS models can help (see Chapter 9). It is always best to be as prevention-oriented as possible and limit reactive responding.

Some specific tactics include developing self-care plans, promoting good communication with the treatment team, and recognizing early signs of caregiver burnout. Trauma-focused CBT aims to improve caregivers' ability to be more proactive and calmer in the critical situations that arise, but also to reduce the caregivers' anxiety and trauma symptoms overall. If effective, these approaches will interrupt the transactionally produced cycle of stress noted above and promote more positive outcomes for both the caregiver and the person for whom care is provided.

Mindfulness

Singh and colleagues found that teaching caregivers to use mindfulness approaches impacts the presentations of the people for whom they provide care, including when there are challenging behaviors. Mindfulness techniques help the caregiver be more present and better able to regulate themselves as they respond to challenges. Harmful and unpleasant reactions to potentially trauma-inducing events, such as people with autism or other forms of IDD engaging in aggressive acts, may be reduced for both caregivers and the person for whom care is provided by using techniques like mindfulness (Singh et al., 2013).

Caregivers may begin the use of mindfulness by thinking about what aspects of caregiving they experience as most difficult or worrisome, such as being fearful of getting hurt or just becoming very tired. It is helpful to "normalize" these feelings and fears and recognize that almost anyone is going to experience stress doing what you, the caregiver, are doing. These are expected human responses. Note that a critical goal of mindfulness is to help the person be in the moment, to be tuned in and "present" within the current moment in time. It has also been suggested that mindfulness promotes a nonjudgmental perspective, including not judging yourself. Showing kindness to the people for whom you may be providing care is, of course, vital, but also showing the same kindness to yourself helps everyone. Ultimately, the use of mindfulness techniques appears to assist people in being calmer and more patient when facing a behavioral crisis. Reflecting on crises in retrospect may also be helped by using this approach, reducing persistent feelings of distress related to recent difficult occurrences (Singh et al., 2013).

Gratitude: helping reduce trauma's impact

The emerging studies have demonstrated that when we share our gratitude, even in the context of a life with traumatic experiences, there may be less stress and suffering as a result. Dang et al. (2025) and Emerson (2015) examined the use of gratitude strategies. Among their study participants who completed gratitude interventions, signs of significantly more positive affect and less negative affect were shown compared to people in four control conditions. This may be yet one more strategy to help caregivers of individuals with ASD and other IDD who engage in repeated acts of aggression. In our work, we often find amazing parents who express gratitude for having a unique child who enriches their lives despite the challenges. In these instances, these parents somehow seem to be coping better with the stress they face.

Positive daily self-affirmations

A final possible tool is daily repetitions of positive affirmations, used to bolster a strong sense of self and to amplify efforts to build resilience in the face of daily challenges (see Creswell et al., 2013). Daily affirmations are simple to use and easy to access as a coping tool. You can find ones that fit your needs, post them on a mirror, and make eye contact with yourself while repeating them. Over time, when done as part of a daily routine, people report improved self-perceptions. Creswell and colleagues note that, when used over time, this will help people become more capable of problem-solving under stress.

Summary

There is help and hope for the caregivers of people with ASD/IDD who engage in aggressive behaviors. Though altering and improving policies and access to needed supports and services is more difficult to accomplish, individuals can seek to employ some of the interventions suggested above. All of us can advocate for greater recognition of the different, chronic stress produced in the process of caring for people with autism and other IDD who present with aggressive behaviors. As described above, interactions are transactional, and helping one of the individuals involved inevitably helps the other (Table 8.1).

Table 8.1 Interventions for caregivers supporting individuals with ASD/IDD who display aggression.

Intervention	Key references	Mechanisms/how it works
Regulation models (Polyvagal-based; rhythmic sensory input)	Porges (2011), Dana (2018), Perry (2021)	• Restores ventral vagal calmness and social engagement • Uses slow exhalation, prosodic voice, soft gaze, and warm facial expressions • Rhythmic movement (walking, rocking, tapping) regulates overactivation
Bilateral stimulation/related therapies	Shapiro (2018), Van der Kolk (2014)	• Uses dual-attention stimulation (eye movements, tapping, tones) • Reduces trauma symptoms by engaging adaptive information processing • Integrates bottom-up and top-down processes
Grounding & sensory regulation strategies	Kalb et al. (2016), Rumball et al. (2020)	• Reduces acute arousal during or after aggressive incidents • Uses sensory grounding (5-4-3-2-1, temperature shifts, deep pressure) • Environmental modifications (noise, lighting, scent) • Stabilizes balance (rooting stance) to reduce panic
EMDR (Eye Movement Desensitization and Reprocessing)	Shapiro (1989), Shapiro (2018), Shapiro (2013), Stekkinger-de Vries et al. (2025)	• Evidence-based trauma processing therapy • Eight-phase protocol (history → reevaluation) • Bilateral stimulation facilitates the processing of traumatic memories • Moves experience from emotional to cognitive understanding

Continued

Table 8.1 Interventions for caregivers supporting individuals with ASD/IDD who display aggression—cont'd

Intervention	Key references	Mechanisms/how it works
Trauma-focused cognitive behavioral therapy (TF-CBT)	Stewart et al. (2017), Kalawatia et al. (2025)	• Addresses trauma-linked thoughts and fears • Interrupts negative caregiver–child stress cycles • Uses grounding, relaxation, cognitive reframing, and safety planning • Reduces caregiver anxiety, shame, and burnout
Mindfulness approaches	Singh et al. (2013)	• Improves present-moment awareness and emotional self-regulation • Reduces reactive fear responses during crises • Encourages a nonjudgmental perspective and self-kindness
Gratitude practices	Dang et al. (2025), Emerson (2015a).	• Increases positive affect; reduces negative mood • Supports meaning-making and resilience • Helps buffer against long-term trauma exposure
Positive daily self-affirmations	Creswell et al. (2013)	• Strengthens positive self-concept and resilience • Counters negative, stress-based self-appraisals • Easy daily strategy to promote emotional stability and confidence

References

Bearss, K., Johnson, C., Smith, T., Lecavalier, L., Swiezy, N., Aman, M., McAdam, D.B., Butter, E., Stillitano, C., Minshawi, N., Sukhodolsky, D.G., Mruzek, D.W., Turner, K., Neal, T., Hallett, V., Mulick, J.A., Green, B., Handen, B., Deng, Y., Dziura, J., Scahill, L., 2015. Effect of parent training vs parent education on behavioral problems in children with autism spectrum disorder: a randomized clinical trial. JAMA. 1538-3598. 313 (15), 1524–1533. Available from: http://jama.jamanetwork.com/article.aspx?articleid=2275445. doi:10.1001/jama.2015.3150.

Beneytez, C., 2023. Intolerance-of-uncertainty and anxiety as serial mediators between emotional dysregulation and repetitive patterns in young people with autism. Res. Autism Spectr. Disord. 1750-9467. 102, 102116. doi:10.1016/j.rasd.2023.102116.

Bentley, T.G.K., Castaneda, P., Constantin, L., Vlisides-Henry, R.D., Moeller, J., 2023. Breathing practices for stress and anxiety reduction: conceptual framework of effective interventions and systematic review of mechanistic evidence. Brain Sci. 13 (12), 1612. doi:10.3390/brainsci13121612.

Charlot, L.R., Hodge, S.M., Holland, A.L., Frazier, J.A., 2022. Psychiatric diagnostic dilemmas among people with intellectual and developmental disabilities. J. Intellect. Disabil. Res. 66 (10), 805–816.

Chireh, B., Essien, S.K., Swerhun, K., D'Arcy, C., Acharibasam, J.W., 2025. Workplace stressors and mental health outcomes among personal support workers: a systematic review. Int. J. Nurs. Stud. 0020-7489. 168. Available from: https://www.sciencedirect.com/science/journal/00207489. doi:10.1016/j.ijnurstu.2025.105093.

Coulman, E., Gore, N., Moody, G., Wright, M., Segrott, J., Gillespie, D., Petrou, S., Lugg-Widger, F., Kim, S., Bradshaw, J., McNamara, R., Jahoda, A., Lindsay, G., Shurlock, J., Totsika, V., Stanford, C., Flynn, S., Carter, A., Barlow, C., Hastings, R.P., 2021. early positive approaches to support (E-PAtS) for families of young children with intellectual disability: a feasibility randomised controlled trial. Front. Psychiatry. 1664-0640. 12. Available from: http://www.frontiersin.org/Psychiatry. doi:10.3389/fpsyt.2021.729129.

Creswell, J.D., Dutcher, J.M., Klein, W.M.P., Harris, P.R., Levine, J.M., 2013. Self-affirmation improves problem-solving under stress. PLoS ONE. 1932-6203. 8 (5). Available from: http://www.plosone.org/article/fetchObjectAttachment.action?uri=info%3Adoi%2F10.1371%2Fjournal.pone.0062593&representation=PDF. doi:10.1371/journal.pone.0062593.

Cuzzocrea, F., Murdaca, A.M., Costa, S., Filippello, P., Larcan, R., 2016. Parental stress, coping strategies and social support in families of children with a disability. Child Care Pract. 1476-489X. 22 (1), 3–19. Available from: http://www.tandf.co.uk/journals/titles/13575279.asp. doi:10.1080/13575279.2015.1064357.

Dana, D., Ed. 2018. Clinical Applications of the Polyvagal Theory: The Emergence of Polyvagal-informed Therapies. W.W. Norton.

Dang, A.V., Coles, N.A., Oishi, S., McCullough, M.E., 2025. The efficacy of seven gratitude interventions for promoting subjective well-being. J. Posit. Psychol., 1–19.

Emerson, D., 2015a. Trauma-sensitive Yoga in Therapy: Bringing the Body into Treatment, W.W. Norton & Company.

Emerson, D., 2015b. Trauma-sensitive Yoga in Therapy: Bringing the Body into Treatment, W.W. Norton & Company.

Giroux, C., Ahlers, D., Miawotoe, A., 2023. Polyvagal approaches: scientifically questionable but useful in practice. J. Psychiatry Reform 10 (11).

Grossman, P., 2023. Fundamental challenges and likely refutations of the five basic premises of the polyvagal theory. Biol. Psychol. 0301-0511. 180, 108589. doi:10.1016/j.biopsycho.2023.108589.

Guo, Y., He, H., Lan, J., 2025. The effectiveness of acceptance and commitment therapy on parental stress in parents of special children: a meta-analysis. Child Adolesc. Psychiatry Ment. Health 19(1), 80. doi:10.1186/s13034-025-00944-y. PMID: 40691592; PMCID: PMC12278553.

Hassiotis, A., Walsh, A., Budgett, J., Harrison, I., Jones, R., Morant, N., Courtenay, K., Crossey, E.V., Hall, I., Romeo, R., Taggart, L.G., Langdon, P.E., Ratti, V., Kirchner, V., Lloyd-Evans, B., 2020. Intensive support for adults with intellectual disability and behaviours that challenge: a survey of provision and service typologies in England. BJPsych Open. 2056-4724. 6 (2). doi:10.1192/bjo.2020.2.

Hastings, R.P., Taunt, H.M., 2002. Positive perceptions in families of children with developmental disabilities. Am. J. Ment. Retard. 107 (2), 116–156. doi:10.1352/0895-8017(2002)107<0116:PPIFOC>2.0.CO;2.

Herman, J.L., 2015. Trauma and Recovery: The Aftermath of Violence—from Domestic Abuse to Political Terror (Rev. Ed.), Basic Books.

Kalawatia, M., Lucke-Wold, B., Mehrunkar, A., 2025. Supporting parents in autism care. World J. Psychiatry. 2220-3206. 15 (5). doi:10.5498/wjp.v15.i5.103575.

Kalb, L.G., Beasley, J., Klein, A., Hinton, J., Charlot, L., 2016. Psychiatric hospitalisation among individuals with intellectual disability referred to the START crisis intervention and prevention program. J. Intellect. Disabil. Res. 0964-2633. 60 (12), 1153–1164. doi:10.1111/jir.12330.

Levine, P.A., 2010. In an Unspoken Voice: How the Body Releases Trauma and Restores Goodness, North Atlantic Books.

Li, S.N., Chien, W.T., Lam, S.K.K., Chen, Z.Y., Ma, X., 2024. Effectiveness of parent-focused interventions for improving the mental health of parents and their children with autism spectrum disorder: a systematic review and meta-analysis. Res. Autism Spectr. Disord. 1878-0237. 114. Available from: https://www.sciencedirect.com/science/journal/17509467. doi:10.1016/j.rasd.2024.102389.

Lunsky, Y., Robinson, S., Blinkhorn, A., Ouellette-Kuntz, H., 2017. Parents of adults with intellectual and developmental disabilities (IDD) and compound caregiving responsibilities. J. Child Fam. Stud. 1062-1024. 26 (5), 1374–1379. doi:10.1007/s10826-017-0656-1.

Miller, O., Dobson, O., Casey, T., Newnam, S., 2025. Work-related violence interventions in the disability sector: a systematic review and systems mapping exercise. Saf. Sci. 0925-7535. 184, 106765. doi:10.1016/j.ssci.2024.106765.

Murphy, F., Nasa, A., Cullinane, D., Raajakesary, K., Gazzaz, A., Sooknarine, V., Haines, M., Roman, E., Kelly, L., O'Neill, A., Cannon, M., Roddy, D.W., 2022. Childhood trauma, the HPA axis and psychiatric illnesses: a targeted literature synthesis. Front. Psychiatry. 1664-0640. 13. Available from: http://www.frontiersin.org/Psychiatry. doi:10.3389/fpsyt.2022.748372.

National Institute for Health and Care Excellence (2015) Challenging Behaviour and Learning Disabilities: Prevention and Interventions for People with Learning Disabilities Whose Behaviour Challenges (NICE Guideline NG11).

NHS England, 2017. Transforming care: Model service specifications (Version 1). Available from: https://www.england.nhs.uk/wp-content/uploads/2017/02/model-service-spec-2017.pdf.

Ogden, P., Minton, K., Pain, C., 2006. Trauma and the Body: A Sensorimotor Approach to Psychotherapy, W. W. Norton.

Peebles, J.L., 1991. Attitude change in staffs trained to work with aggressive clients. In: 25th Annual Meeting of the Association for Advancement of Behavior Therapy, New York, NY.

Peebles', 1992 Attitude Change in Staffs Trained to Work with Aggressive Clients (Dissertation) from Boston College.

Perry, B.D., Winfrey, O., 2021. What Happened to You? Conversations on Trauma, Resilience, and Healing, Flatiron Books.

Porges, S.W., 2011. The Polyvagal Theory: Neurophysiological Foundations of Emotions, Attachment, Communication, and Self-regulation, W. W. Norton.

Reyes-Martín, J., Simó-Pinatella, D., Andrés, A., 2024. Emotional reactions of professionals to challenging behaviors in people with intellectual and developmental disability. Behav. Sci. 2076-328X. 14 (8), 707. doi:10.3390/bs14080707.

Rodriguez, G., Hartley, S.L., Bolt, D., 2019. Transactional relations between parenting stress and child autism symptoms and behavior problems. J. Autism Dev. Disord. 1573-3432. 49 (5), 1887–1898. Available from: www.wkap.nl/journalhome.htm/0162-3257. doi:10.1007/s10803-018-3845-x.

Rumball, F., Happé, F., Grey, N., 2020. Experience of trauma and PTSD symptoms in autistic adults: risk, vulnerability, and caregiver implications. J. Autism Dev. Disord. 50, 4321–4333.

Ryan, C., Bergin, M., Wells, J.S.G., 2021. Work-related stress and well-being of direct care workers in intellectual disability services: a scoping review of the literature. Int. J. Dev. Disabil. 2047-3869. 67 (1), 1–22. doi:10.1080/20473869.2019.1582907.

Sánchez Amate, J.J., Luque de la Rosa, A., 2024. The effect of autism spectrum disorder on family mental health: challenges, emotional impact, and coping strategies. Brain Sci. 2076-3425. 14 (11). Available from: http://www.mdpi.com/journal/brainsci. doi:10.3390/brainsci14111116.

Sapolsky, R.M. (2004). Why Zebras Don't Get Ulcers, 3rd. Owl Books/Henry Holt and Company.

Shapiro, F., 1989. Eye movement desensitization: a new treatment for post-traumatic stress disorder. J. Behav. Ther. Exp. Psychiatry. 0005-7916. 20 (3), 211–217. doi:10.1016/0005-7916(89)90025-6.

Shapiro, F., 2013. Getting past Your Past: Take Control of Your Life with Self-help Techniques from EMDR Therapy. Rodale Press, New York, NY, pp. 352.

Shapiro, F. (2018). Eye Movement Desensitization and Reprocessing (EMDR) Therapy: Basic Principles, Protocols, and Procedures, 3rd. Guilford Press.

Singh, N.N., Lancioni, G.E., Winton, A.S.W., 2013. Training in mindful caregiving transfers to parent–child interactions. J. Child Fam. Stud. 22 (6), 739–746. doi:10.1007/s10826-012-9629-7.

Stekkinger-de Vries, A., van Ee, E., Verschuur, R., Mevissen, L., 2025. EMDR-therapy for child-related PTSD in parents of adolescents with autism spectrum disorder and severe emotional dysregulation: a multiple baseline evaluation. Eur. J. Psychotraumatol. 2000-8066. 16 (1). doi:10.1080/20008066.2025.2555792.

Stewart, M., Knight, T., McGillivray, J., Forbes, D., Austin, D.W., 2017. Through a trauma-based lens: a qualitative analysis of the experience of parenting a child with an autism spectrum disorder. J. Intellect. Dev. Disabil. 1469-9532. 42 (3), 212–222. Available from: www.tandf.co.uk/journals/titles/1326978X.asp. doi:10.3109/13668250.2016.1232379.

Teicher, M.H., Samson, J.A., Anderson, C.M., Ohashi, K., 2016. The effects of childhood maltreatment on brain structure, function and connectivity. Nat. Rev. Neurosci. 1471-0048. 17 (10), 652–666. Available from: http://www.nature.com/nrn/. doi:10.1038/nrn.2016.111.

van der Kolk, B.A., 2014. The Body Keeps the Score: Brain, Mind, and Body in the Healing of Trauma, Viking.

CHAPTER 9

Developmental contextual assessment and treatment: A positive behavior support (PBS) approach with added features

Lauren Charlot
UMass Chan Medical School, Worcester, MA, United States

Understanding aggression: A developmental contextual assessment and treatment approach

While there are other interventions that have an evidence base for managing behaviors that challenge, e.g., anger, here we focus on an adapted version of positive behavior support (PBS), which has been borne out of clinical practice and which has, at its core, the support of people who display aggressive behavior. This chapter discusses an approach that builds on PBS. Early on, and in subsequent iterations, PBS models have embraced the notion that aggressive behavior can only be understood by looking at multiple factors that impact its occurrence among people with autism and other IDD.

"Positive behavior support (PBS) is defined as an approach for dealing with problem behavior that focuses on the remediation of deficient contexts (such as environmental conditions and/or behavioral repertoires) that by functional assessment are documented to be the source of the problem" (Carr, 1999).

Hunter and Gardner described what was called a multimodal functional model, in which they built on traditional "diagnose and treat" approaches, expanding on traditional applied behavioral analysis (Hunter et al., 2008). They discuss the need to examine a person's "vulnerabilities" in addition to looking at *setting events* and *triggering* events. Essentially, this model suggests that basic functional assessments may not be enough to understand and develop effective interventions for concerns like aggressive acts, but especially to see how these problems relate to psychiatric conditions. The approach described here provides a similar framework, but with the

Aggressive Behavior of People with Autism and other Intellectual and Developmental Disabilities
https://doi.org/10.1016/B978-0-443-26552-5.00011-4

added attention to developmental features that likely impact how people experience stress. Assessment should include consideration of developmental influences on behavior for several reasons (Lehman et al., 2017). Developmental features modify acute mental illness symptoms so that, for example, anxiety expressed by a young person will usually be manifested differently than how it is displayed by an adult. For people with ASD and other IDD, the person's developmental profile may be more important than chronological age in understanding the roots of aggressive behaviors. Using a developmental perspective helps distinguish challenges due primarily to complications of developmental difficulties from those representing psychiatric phenomena. The suggestion here is that adding a *developmental perspective* will improve identification of pathways to aggressive behavior in this population. The other added feature is an emphasis on studying and identifying contextual influences on the probability of aggressive behavior. Contextual behavioral assessment and treatment planning, combined with a developmental perspective, can improve outcomes. Contexts to be examined are individually identified and can include a wide array of factors described in more detail below. Prior work pointing out the benefit of contextual assessment was a starting point for the approach suggested here (Cale et al., 2009; Carr and Horner, 1991).

The proposed *Developmental Contextual Assessment and Treatment* approach considers contextual influences as a significant area of focus in behavioral assessment and treatment and provides a bridge to multidisciplinary, multimodal interventions through its use of a developmental framework. It is primarily different from the usual bio-psycho-social model in the importance placed on understanding a person's developmental profile as a starting point. The suggested approach has not been formally researched, though, as noted above, it is based on models proposed previously, along with the author's clinical experiences. It is suggested here that behavioral health case formulations for people with autism and other IDD who present with possible psychiatric conditions and aggressive behaviors are enhanced when using an individualized developmental profile with contextual assessments. The developmental-contextual approach informs a *developmental*-bio-psychosocial assessment and treatment plan, which, in turn, is further defined by studying contextual influences. With the recommended tools, an effective positive behavior plan that addresses the use of multiple modalities can be developed.

Case example of contextual assessment

Calculating the probability of aggressive Behavior can be based on several suspected contextual influences, as was done for a gentleman with autism, Mr. N, seen by author Charlot. The caregivers collected data and shared it for a contextual assessment. The Table below shows the percentage of occurrences of aggressive behavior in the setting of each of the contextual elements listed. Staff were certain that family visits and poor sleep were the culprits. This young man, aged 32, had never slept through the night since childhood, according to his parents. It was pointed out to the treating physician, who had diagnosed him as having Bipolar Disorder, that it might not be considered a new symptom explaining the onset of aggressive behavior when he was an adult. Data showed that Mr. N was no more likely to engage in an aggressive act when he slept only a couple of hours than when he slept through the whole night. Data also demonstrated that aggressive behavior was more probable when new staff were being trained and when he wanted to go home. A new plan for staff orientation was developed, pairing trainees with caregivers he liked a lot (his "preferred staff"). In addition, his functional communication challenges were a developmental component of his stress that was attenuated over time. He was helped to learn to use words and sign language to tell people he was finished with activities when he wanted to leave. Previously, his aggressive acts were his primary way of getting taken out of situations (for him, these were crowded or loud settings) and getting new staff to leave. Other data showed that aggressive behavior was less probable when he was able to go for a coffee before his daily activity, so this became a part of his daily routine. He began to use more functional communication and eventually had medications scaled back, with all agreeing that he did not show the typical sleep challenges, episodic patterns of more acute challenges characteristic of Bipolar Disorder. Rather, he seemed much more troubled by struggles associated with his unique developmental profile. See additional details regarding a developmental approach in Chapter 2.

Example analysis of contexts for aggressive behaviors.

n = 50 incidents	Freq	Per cent
New staff	22	44%
Family visit	3	6%
Crowded YMCA	8	16%
Poor sleep	4	8%
Constipation	13	25%

Contextual influences

Some behavioral treatment specialists emphasize the identification of "maintaining" variables (attention, escape, automatic) as key to understanding ongoing incidents of aggressive behavior. The functional assessments aimed at finding out what factors maintain a given behavior are very important and need to be evaluated when trying to better understand incidents of aggressive behavior. A contextual functional assessment and treatment planning approach includes looking at maintaining variables. However, it focuses more on determining the probability of a given behavior under varied conditions or contexts. As suggested by Carr, this can be most helpful for developing interventions for the purpose of *prevention,* as described with the case example above. In the case of Mr. N, functional communication was taught, and activities were added that were found to be associated with feeling happy and having a lower probability of aggressive behavior. Understanding how these struggles for Mr. N were also associated with his profile of ASD-related developmental features informed a new psychiatric case formulation and helped caregivers understand him in new ways. The focus on these factors expands our understanding of the consequent events that reinforce the behavior of concern. Mr. N was at times "escape-motivated." Learning why certain experiences caused him anxiety (the contexts of the events) led to more effective interventions that reduced his aggressive behavior for the first time in several years.

Using a contextual assessment does not mean that consequent events are not considered or measured. It is just that their meaning may be further explained in ways that inform interventions that may prevent behaviors of concern. Contexts with possible influence, as suggested above, can include many different factors. It is helpful to first look at the time of day and then examine variables such as after or before mealtimes, during caregiver shift changes, during times when there are no activities, during times when there are specific activities, when on a community trip, when a family member is visiting, when at school or home, when the environment is loud, etc. (see additional examples below). Lack of sleep, family contacts, specific settings for outings, and medical conditions for which data were taken for Mr. N provide other examples of contexts one can track. In all cases, it is advised to track sleep for both day and night hours. Many people are helped by tracking fluid intake, bowel movements, and signs of fatigue or movement difficulties. Some contexts are more remote, such as when an important event is going to occur soon or has occurred recently, such as an annual meeting at which important things will be discussed. These things may be impacting the person in ways that alter the risk for aggressive behavior, despite them not being visible at or near the time of an incident of aggressive behavior. The contexts that may be studied to determine if there is evidence of them impacting outcomes are first identified through talking with the individual and their caregivers, as well as looking through records that address past mood, mental status, and behavior. Though identifying and *maintaining* variables is important, as Carr and others suggested, attention to contexts may be more key to informing more preventive interventions (Carr and Horner, 1991; Ghojoghi et al., 2024; Verhoeven and Egger, 2014).

The proposed *Developmental-Contextual Assessment and Treatment* data tracking tools are designed to identify the most frequent *contexts* associated with *both* positive and negative mental status, mood, and behavior. If implemented successfully, the findings may be used to modify these influences. Aggressive and other significant challenging behaviors are seen as representing a *final common pathway* for frustration and distress in people who may have developmentally associated limitations in their means of expressing themselves. In one case, for example, a person who was aggressively hitting herself in the face was described, based on a usual functional assessment, as being reinforced by attention. A plan was developed and implemented to reduce attention when this behavior occurred, while more attention was delivered when it was absent. However, because a clinical consultant insisted on further assessment, it was discovered that she had an infected tooth. It was important to understand that this was the cause of her self-directed aggression, even though it was gaining staff attention. While knowing she was seeking caregiver attention was relevant, determining that a medical factor was influencing its occurrence in this instance was far more crucial. She had a long history of self-injurious behavior but had not been engaging in it for some time, and she was showing an overall positive response to medication reductions. Understanding why she was engaging in this behavior at the time was critical to resolving her distress, which, in turn, stopped her self-injurious acts. In her case, when facial pain was present, the probability of an aggressive act was much higher than when it was not. In many instances, pain lowers the threshold for reacting to small "triggers" by engaging in an aggressive act. For this individual, she disliked being asked to brush her teeth but usually cooperated with encouragement. If she were experiencing pain, the request was much more likely to result in her becoming self-injurious. Because of this event, her PBS plan listed checking for any possible physical distress as a first step when aggressive behavior reemerged or worsened. Another clue in her case was a history of refusing dental care that had not yet been resolved.

In many systems of care, teams and clinicians see their work as aimed at reducing aggression or controlling it. This is true many times, even when there are also psychiatric diagnoses. Negative behaviors of concern are often tracked, while signs and symptoms of the identified psychiatric disorder may not be.

Case Example regarding what data we seek when supporting people with cooccurring psychiatric conditions? Ms. X was a 26-year-old woman diagnosed with moderate ID, ASD, and bipolar disorder, living in a group residence. Data were collected on the frequency of aggressive acts and was provided to her new psychiatrist. However, no data had been collected on occurrences of symptoms of mania or depression, even though the program was mandated to track hours of sleep at night. Ms. X was treated with four psychotropic medications for her emotional and behavioral challenges. The assessment of treatment efficacy rested entirely on rates of aggressive behavior. If she exhibited fewer aggressive behaviors, the mood disorder was described as improving. Whenever she experienced an uptick in rates of aggressive behaviors, her medication was increased. If she slept little at night, her medication was increased, and at times, additional medications were added to target sleep. Using a newly designed context-focused data sheet, caregivers tracked sleep (both day and night), signs of positive mood (smiling, laughing), crying, extreme overactivity, the activity she was engaged in, complaints of stomachaches, hand-wringing, and incidents of aggressive behavior. For her, aggressive behavior included slapping and kicking caregivers, usually while screaming. Data were collected for 24 h at hourly intervals, noting whether the listed items were present or not, over a 4-week period. Please see example data sheets. The data showed that the probability of aggressive acts was greatest at each mealtime and during shift changes, when it was noted to be loud. As it turned out, she did have a few hours of sleep at night, but she generally napped every day. Her mood was specific to upsets, and the probability of a positive mood was about 80% of intervals. Only crying and stomachache complaints were associated with a high probability of an aggressive act. Further medical assessment showed she had gastroesophageal reflux (GERD), and when treated, aggressive behavior was reduced. She was also noted to be less likely to be aggressive when she was engaged in an activity, and intervals without an activity were associated with more risk. She only seemed overactive in the immediate context of an outburst. Each of these findings about the probability of seeming well or seeming distressed and unwell served to inform a new plan of approach that was effective in reducing her aggressive behavior and reduced reliance on psychotropics for bipolar disorder, since she lacked a decreased need for sleep and was not persistently overly energized. It was also evident from learning who she was when well and understanding her developmental strengths and challenges that she enjoyed being busy doing puzzles and had a sound sensitivity. She spoke well enough to tell caregivers her stomach hurt, but this had previously been viewed as an attention-seeking measure.

Case example

A 28-year-old female diagnosed with psychosis and spectrum disorder was placed in a facility. The following is an initial functional contextual (behavior) assessment based on interviews, records reviews, and a review of data from 24-h hourly interval or CAT data-tracking sheets.

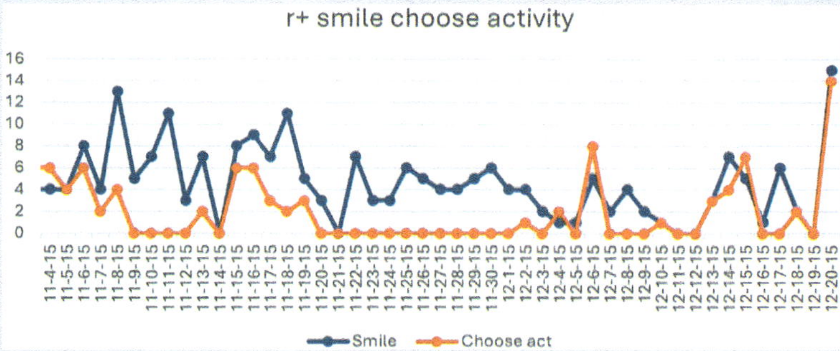

With some exceptions, there is a trend suggesting key interventions/targets will be to improve her mood and activity engagement levels. A third factor likely acts as a moderator (see below). Being active is associated with being happier. Being happier is associated with reduced challenging behaviors (see below).

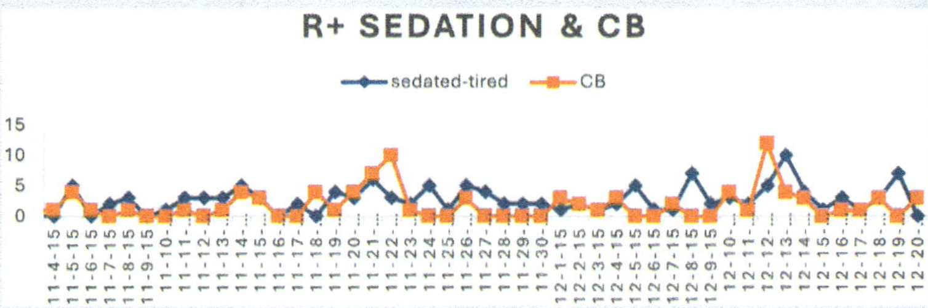

The recommended data system provides the opportunity to collect data regarding observable signs or factors suspected to be associated with positive mood and behavior, as well as the absence of aggressive behavior. There are usually some observable signs of psychiatric symptoms. Eating, daytime sleeping, activity refusals, crying, and many other similar phenomena can be tracked on a 24-h hourly tracking form (example provided below). Selecting what should be tracked is highly individualized, often related to each unique person's developmentally associated manner of expressing or demonstrating various states (rapid vocalizing or rapid speaking, for example,

***** CB = Challenging Behaviors: kicks, hits,** etc. There is a significant relationship between this person being tired, daytime sleep, signs of sedation, and the probability of CBs. Other factors have an influence, but a major treatment target needs to be reducing sedation.

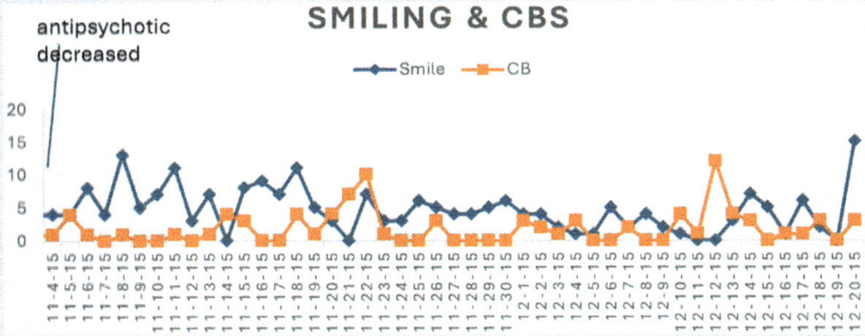

Periods of positive mood may be associated with a lower probability of CBs. Improvement was observed after a reduction in antipsychotic medication.

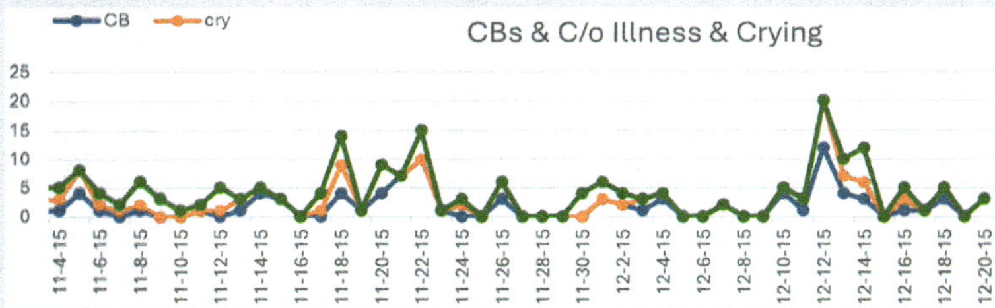

There appears to be a strong association between this person stating she was not feeling well, crying, and engaging in CBs.

going to bed during the day as a sign of fatigue versus telling people, ("I am tired."). The contextual assessment and treatment (CAT) approach uses many basic principles found in traditional Applied Behavioural Analysis (ABA) and Positive Behavior Supports (PBS) models but has an expanded form of assessment to be integrated into a developmental-bio-psychosocial assessment and treatment plan for addressing aggressive behavior.

The primary treatment goal when using a contextual approach for behavioral treatment planning is to help people be *well* rather than simply less aggressive. A desired outcome is, of course, for the person to be less aggressive, but to achieve this by becoming as *well* as possible. This, again, is based on the person's unique developmental profile. As noted, the *Developmental Contextual Assessment and Treatment data* collection includes tracking signs of wellness. When people are helped to be healthy and happy, as well as possible, aggressive acts (and other challenging behaviors) are less likely to occur. On the other hand, when the focus is primarily on reducing aggressive behavior, there may even be a negative quality-of-life outcome for the person. The use of many restrictive reactive approaches can eliminate or reduce aggressive behavior. However, often, the reductions in aggressive behavior remain only as long as the restrictions continue. When asked what to do when a person throws a chair at you, the best advice is to "duck." To prevent chair throwing, however, we need to understand how and why this individual got to the point where throwing a chair at someone became their "go-to" response. This means starting with an understanding of what is usual for this person when well, when the individual was last well for a sustained period, and then what changed since that time. In addition, it is necessary to identify the current contexts in which aggressive behavior is more probable. It is also useful to examine what past and present contexts or factors are present when happiness or wellness is more probable. For example, several patients treated on the specialized psychiatric unit referenced in Charlot et al.'s 2010 paper demonstrated the problem with a "*stop the aggression*" treatment focus. On admission, these individuals were described as recently having a major reduction in aggressive acts. Unfortunately, these same patients were barely awake for much of the day and had recent psychotropic medication additions and dose increases. The patients demonstrated not only reduced aggressive behavior but reductions in almost *any* behavior. Clearly, less aggression should not be the primary goal, whereas wellness can and should be. Promoting wellness is an important, often effective strategy to reduce aggressive behavior. The CAT approach provides a set of tools and a framework aimed at identifying influences on behavioral outcomes that may be modified or amplified, ideally improving quality of life while also reducing aggressive behavior.

Overview of a developmental–contextual assessment and treatment planning approach

An outline for developing a plan based on the approach described here is provided. The aim is to have a plan that is broad-based and uses varied modalities, as illustrated below. Assessments are designed to determine when, where, with whom, and under what conditions the person seems more distressed and at greater risk for engaging in aggressive behavior, as well as when, where, with whom, and under what circumstances the individual displays more signs of wellness and happiness (thereby rendering aggressive behavior less probable). Additional elements or steps are required and include first understanding who the person is when well, including the person's developmental profile. The aim is to understand if the individual's unique strengths and areas in which they face challenges have impacted how they are presenting now. Aspects of strengths and challenges are derived from one's developmental profile but also from past experiences or contextual influences. Some of these occurred earlier or more remotely, such as experiencing trauma. As described further in Chapter 7, many people have an overall lowered threshold for becoming distressed when they have a history of being traumatized. Early experiences in the family may also have persistent influences. In some cases, a parent has skills in teaching their child with ASD how to functionally communicate better, improving an outcome that might otherwise be more negative. In other instances, youth with autism assault family members who cannot manage the aggressive behavior and they are removed from the family home. These transactions impact the individuals involved in them. A key step in assessment is learning the individual's story—essentially the *history of the presenting problem or illness*. The history of both past care that was effective and interventions that were problematic or unhelpful needs to be sought, if available. A CAT treatment plan includes tactics addressing several areas other than what is assumed to be "learned behavior." A goal is to identify influences that increase the probability of aggressive behavior and use tactics to mitigate their influence. A further aim is to examine factors present when behavior indicative of the person being calm or well is noted, since this provides an opportunity to add more of these events or activities.

Any influences we can identify can be addressed or at least better understood. For example, on our UMass Medical specialty psychiatric unit, the team found people who had a greater probability of aggressive behavior after taking a sedating medication. One gentleman seemed to get aggressive to gain attention, and knowing this was somewhat helpful. Finding out he was specifically more likely to display aggressive behavior after he was observed to be sedated by a specific medication was important. He was seeking people to help with his discomfort; later he was able to say he felt "fuzzy," and this was upsetting. Knowing the context was critical in reducing his aggressive behavior.

The approach suggested is also designed to inform the development of treatment plans that are *practical*. When plans are long and detailed without specific guidance for caregivers, consistent use may be compromised. Please see the example plan below. Tools and methods to determine the most likely factors that influence the occurrences of both positive and negative behaviors, as well as altered mood or mental status, are provided. Informants or caregivers may report aggressive behaviors happening at *any time of day* and *in any setting* (home, school, work, community trips, etc.). This may be true. However, in most instances, the frequency or intensity of aggressive behaviors will be more probable under certain circumstances. As suggested, a more contextual form of assessment helps the clinician identify factors associated with a greater probability of either positive mood and behavior states or aggressive behaviors and agitation.

As noted, many people with autism or other IDD have challenges expressing their distress, even when verbal. Many individuals, especially people with autism, may struggle to understand social cues and the demands of a social context. These and other developmental challenges increase the risk of aggressive behaviors that represent a surface manifestation of underlying distress, discomfort, or frustration. It is important to remember that aggressive behavior is not diagnostically specific and is likely to be displayed under a wide array of circumstances and in association with many different psychiatric phenomena, especially in the population we serve. If aggression is a final common pathway for distress, as stated earlier, our job is to find the pathways involved in each case and even in each situation. When contextual influences are identified, plans may be written to either prevent or mitigate these influences. See the example plan below (Box 9.1).

Developmental-contextual assessment and treatment plans: Step by step

Step 1: Learn about this person's history and functional abilities

After a review of basic identifying data (age, gender, where they live, services, and current psychiatric and medical diagnoses) and the reason for referral, the first step in the proposed model is to understand who this unique person is by examining aspects of his or her personal profile. This review is based on what can be learned about this individual's neurodevelopmental features, as described in Chapter 2.

Learning a person's history can be difficult, especially when treating or consulting to help a person with whom you do not have a long-term relationship. Also, if a person cannot reliably self-report the needed information, it will be necessary to rely on interviews with parents, other family

Box 9.1
Example of an informed plan using the developmental contextual approach

Overview: AAAAAA is a XX-year-old male with XX ID, carrying psychiatric diagnoses of Anxiety Disorder and ASD, as well as a medical diagnosis of constipation.
Referred due to: Increase in "meltdowns" and anxiety.
Coming from: Family home.
Communication needs: May not understand some things and won't ask; needs reminders and the use of communication cards.
Daily care needs: Only needs reminders and setup; does everything himself.
Recent stressors: His mother has been ill, and he has had a lot of changes at school.
Strengths: Humor, Warm, and caring.
Challenges:
Aggression to others—Might slap or kick if people are trying to make him move when he is too upset (f = 3 x's/mo)
Aggression to self—bangs his head on the floor if sitting on the floor and people try to move him (f = 6 x's mo)
Aggression to property—**punches** the wall (f = 3 x's per week)
Leaves supervision and is unsafe. NO. Suicidal? NO.
Meltdowns: Main concern—he will drop to the floor, yell out swears, scream high-pitched, and if it goes on for more than 2 minutes, it becomes very hard to help him calm down; **Threats:** NO; Name-calling, swearing, etc.: YES; Functional decline: NO.
Likes (What will put a smile on his or her face; things that can be done here)

- Recognition when doing well; loves to help and please others
- Playing video games
- Basketball
- Walks
- Cats and dogs, animals

Need to know: Special important things to know, like don't say "no"; prefers to shower only in the morning; takes medication with juice; family visits daily, etc.
Environmental concerns, 1:1 needed:

- 1:1 constant observation. Point plan (he earns points that can be turned in for rewards).

How to Develop rapport—help a person feel safe and cared for: Play a game with him, and ask him about his favorite games, the Lakers, or the family cat
Setting Events:
Very significant challenges to short-term memory (learns a coping skill, forgets it when upset).

- Despite how much he seems to know, he still needs reminders regarding what comes next, each daily event, reassuring him that help is always nearby. Use visuals—see attached schedule.

Box 9.1
Example of an informed plan using the developmental contextual approach—cont't

Speaks well but seems more able to plan than he actually is; challenges to social understanding.

- Don't expect him to pick up on subtle cues; help him understand what is expected gently and kindly.
- He might tell someone to get away if they sit too close to him; help him ask for space proactively.
- AAAA, say "Space, please." Model this and praise him when he does it.
- Help him create a comfortable space in groups before an issue arises.

Lacks theory of mind; does not understand competing perspectives.

- Please know he is not being mean; he does not realize that his actions could upset others.
- Help people not to judge when he misunderstands. (We will be teaching him skills, but this will take time.)

Anxious, especially about new things, new people, and changes in routines.

- Give him a lot of reassurance that we are here to help him, he is safe, and things will be OK.
- Remind him it's OK if he needs any help; just tell us—remind him we will help every 2 h.

Trauma history: was bullied severely in school.

- Immediately intervene if others say anything "mean" to him or tease him.
- Help him separate.
- Tell him that we all know he is a great person, then distract him with an activity.

Sensory sensitivities to noise and people being too close.

- Visit sensory room 4 × daily (see schedule).
- He loves holding the weighted cat.
- Use a rocking chair, a weighted stuffed animal, and listen to a favorite tune.

Repeats and perseverates

- Provide opportunities to do things enjoyed between other activities.
- Validate. Don't say, "Don't talk about that."
- Distract—e.g., hey, can you come see what I have here for you.

Reads and writes but may "think in pictures."

- See schedule, use pictures—communication cards (are at the nursing station), use First-Then cards, "help."

Difficulty organizing his own time, while really struggling when not engaged.

- See schedule (Picture schedule is attached.)

Has difficulty with sleep—naps during the day and doesn't stay asleep at night; recently tired all the time—may lower the threshold for reacting to small triggering events.

Continued

Box 9.1
Example of an informed plan using the developmental contextual approach—cont't

- He needs to stay up during the day, use humor, joke with him—dribble out of the room.
- One hour before bedtime (10 p.m.), read from his chosen magazine or book (quiet things down, use low lights).
- Team is working on reducing sedative medications administered during the day.

Constipation—more likely to escalate when he is not feeling well.

- Bowel charting
- Alert the nurse if he isn't having a large, soft BM at least every 2 days, or if he has hard BMs, or if he has a stomachache.

Disappointed in himself, he had an incident earlier in the day.

- Provide extra reassurances
- Help him keep busy
- If he brings it up, tell him that's in the past.

Triggers

Perceives someone is telling her what to do and not asking her.

- Always provide some choice, i.e., do you want to go shower now or after dinner?
- Ask politely and do not tell him what to do—it's time for the next group. Will you walk with me?

Loud noises—someone else is upset, especially if they are yelling or crying.

- Move him out of the area if it's getting loud.

Ambulances, sirens

- Use his noisecancelling headphones

Being forced to sit too close to people she doesn't know well.

- Help him get seated where he won't have this problem.

Someone is saying negative things to her about him.

- Get him away from this situation; provide reassurances.

Unexpected change in routine

- Follow plan and schedule
- See a separate plan for fun "surprises."

Staff people are being stern and authoritarian approaches.

- Use a kind gentle voice
- Show him, through posture, voice tone, and expression, that everything is OK.

Cues/Early Warning Signs

Gets quiet—Paces—Tense facial expression, furrowed brow

- Ask him if he is OK. Is anything bothering you?
- Offer to help, but also try to encourage him to use his "I need some help" card.
- Ask if he wants to talk in the sensory room or his room—you choose.
- Want to take a walk?

Box 9.1
Example of an informed plan using the developmental contextual approach—cont't

Reactive Strategies: Hierarchy of Responses (Goal to minimize negative outcomes, provide graceful ways out.)

If all of the above fail to help and he has put himself on the floor, started to yell:

- Back off, but remind him that you want to help.
- Direct others to another area.
- Tell him you are very sorry that he is having a tough time.
- Reapproach him, offering the calming activities listed above every few minutes.
- Switch off staff if he isn't responding—he tends to wind down after 10 min, but may remain upset.

members, or caregivers who know the patient well. Even this is not always possible, and then we do the best we can to learn about how this unique individual thinks, learns, and expresses themselves. In all cases, this must be integrated with whatever historical records are available.

Critical areas

Factors that help us understand how the person functions in his or her day-to-day life:

- Issues identified from birth and early development
- The person's ability to communicate basic things (wants and needs) in relation to what would be expected at their age
- Does the person speak using only single words, phrases, or full sentences?
- Can a person read? Write?
- Ability to communicate reliably about their own personal and medical history
- Ability to understand what others are telling them verbally
- Ability to follow steps, such as tasks requiring 1, 2, or 3 steps, verbally conveyed
- The person's ability to remember things across the day, week, and longer
- Can he or she understand a calendar? Judge the passage of time?
- Able to do daily self-care independently, or requiring help?
- What are the usual patterns of eating, sleeping, and physical activity when healthy and well?
- Is there a copy of a recent physical exam and medication administration record?

- Are there any formal assessments completed that can be reviewed, such as neuropsychological testing, cognitive or IQ testing, assessment of adaptive behavior, or autism assessments?
- Other assessments help as well, such as Speech and Language, Physical Therapy, and Occupational Therapy notes.
- Annual Plans and Individual Education Plans

As noted below, it may be necessary to use current reports and historical documents to supplement patient and informant interviews. The history we obtain will be our best summary of contextual influences on the person's past. These often sustain a significant influence on future aggressive behaviors.

It is critical to know who the most important people in this person's life are.

It is helpful to have a timeline, as described in Chapter 5. In this, we may learn a past pattern where certain contexts were associated with more or less aggressive behavior.

Step 2: Have the person and/or caregivers who know them well describe daily life

It can be helpful to start by asking people what brings them joy and what makes them smile. Ask about family if it isn't upsetting (based on history and the information above). Are they able to have the social experiences they want? Are they able to do the things they want to do? Are they using their strengths to pursue the goals they have for themselves? Where are they living, and what kind of situation is that?

In many instances, people have little to do all day, for example, when aggressive behavior has caused them to have day services discontinued, or when none will accept them due to the aggressive behavior. It is important to know what this person is doing and what a usual day is like.

Step 3: Identify and define behaviors of concern and behaviors that indicate positive states

It is critical that people can identify the individual's strengths and behaviors that challenge. When describing any "targets" for data collection and treatment plans, ask reporters to describe what was said, done, seen, or heard rather than relying on past labels. Use strong, clear operational definitions.

A staff reported:

Jose was manic.

When asked to describe what mania looks like, if the clinician were observing, what would she see? Staff then reported:

Jose was pacing back and forth in the hallway.

Step 4: Assessment tactics in plans based on contextual assessment

There are different ways to gather data. The method proposed here requires that there are people who can fill out a 24-hour data sheet. This tool covers a full day, with caregivers asked to mark YES in the space provided if the item listed was observed or heard, and NO if it was not. Data on frequency could be entered if deemed important. The vertical side of the tool has all the items being tracked which should include operationally defined representations for data that will be helpful regarding the individual. The horizontal axis has the hourly intervals starting with 7am–8am and then having all 24 hours. The form can be fit to a single page and uses a grid. Then it is easy to add up intervals and see what occurrences track together or follow one another across the day. This can help direct the design of interventions. It is often sufficient to calculate a summary using hourly interval data for a day, week, or longer, with just the percentage of time the item was noted. For example, if the person smiles and one can see this when he or she is in a positive mood and is not aggressive, data can be collected for how many hours this behavior was observed. If the person is described to a prescriber as irritable much of the time, data might show that this occurs less often than it feels to caregivers. This can be due to the intensity and significance of the negative affect especially if it is associated with aggressive acts. In many cases, caregivers assume that key factors provoking aggression are the ones they observe.

When the negative, targeted, challenging behaviors are documented in data systems, it is helpful to count episodes in many instances. If a person tends to display a cluster of various signs of upset together most of the time (yells, hits people, bangs on walls), it is better to count these as one episode rather than counting individual items (e.g., hit 2 times, banged the wall 10 times, yelled 12 times), especially when knowing these always include the same or similar 3–4 behaviors. Incident reports can then provide a more detailed description of the event. ABC, or antecedent-behavior-consequence, forms should also be completed if not clearly contained in incident reports when there is aggressive behavior. ABCs, if done well and consistently, can provide or reinforce contextual data collected on daily, hourly interval sheets (shown below).

It is important to have an item tracked for the person's activity and whether they were either engaged in a task activity or having "downtime." In addition, data regarding any complaints of feeling ill, refusing important things, being slowed down or overactive, appearing tired with droopy eyes, yawning, gait problems, or falling down, etc., may all be tracked if indicated by the history gathering and recent concerns. It may also help to note if the person is staying in their room alone or is out with housemates, staff, or family.

Another helpful feature is that one can calculate how often things like smiling occurred during the same hour as other varied occurrences, i.e., family visiting, a snack being provided, or particular situations taking place, such as assaults on staff or other disruptions by housemates. In group care, shift changes may coincide with aggressive behavior more often than at any other time of day. For others, aggressive behavior may be more probable an hour after medications are given in the morning. Again, the goal here is to identify the probability of various mood states and behaviors (especially aggressive behavior) in the context of various factors, telling us where, when, with whom, and under what conditions the item of interest is more probable.

The data form shared includes 24-h sleep data. This is important. If the person naps, even for a few minutes, this should be documented on the 24-h data sheet. Many people assessed by my team were described as having poor sleep at night, which was reported to their MD, who then prescribed medicine to help them sleep better, only for it to be discovered much later that the person slept or was inactive during the day. This is a frequent issue because caregivers often only report sleep data for overnight hours. The CAT data form includes sleep tracking for everyone needing it for 24-h.

Step 5: Develop data-informed, hypothesis-based treatment plans

Using information from the above steps, we can develop hypotheses about some of the critical pathways to aggressive behavior and other behaviors that challenge, understand what to teach and how to teach in the way of skills needed that we can realistically expect the person to learn (using the methods in response to which we know they can learn), design programs with positive strength-based activities and settings in place that the person wants, and develop step-by-step prevention interventions addressing setting events, triggers, and ways to respond to early warning signs that often occur prior to negative incidents. The reaction to the identified negative target event should be the very last step in a plan and be the least elaborated. This is so in CAT plans, designed for caregivers, to guide caregivers to focus on helping people be well. Throughout this, the most critical step is to speak with, observe, and learn from the patient themselves to the greatest extent possible while respecting the true meaning of being "person-centered." At times, this is translated as a need to shape a person's actions to be what we think is more "appropriate" behavior. Whatever is sought needs to be safe for the person and those he or she needs in their life. However, being person-centered means we do all we can to support the individual in getting what they want in their lives. We all seek to have the things we want in our lives. People with ASD and other IDD deserve this every bit as much as anyone else (Table 9.1).

Table 9.1 Components of a developmental–contextual plan: Quick reference.

Overview	Characterization of IDD, psychiatric diagnoses, and medical diagnoses
Referred due to	What exactly brings a person to clinical attention?
Coming from	Where are they living, with whom, and what day services or activities are in place? List any recent moves and changes (see RSQ notes below).
Describe person's neurodevelopmental profile	If using standard testing—look at variability across domains, such as when IQ is in the average range but visuospatial skills are much greater or much lower than verbally mediated ones, or when processing speed or working memory are relatively impaired in contrast to other domains.
Personal strengths	Here is a list of areas where the person demonstrates their most well-developed skills and passions, such as being musical, artistic, creative, good with animal care, organized and able to sort things, capable of completing complex puzzles, excellent at using software for games, and passionate about rap, which they perform well.
Strengths and challenges	Examine as detailedly as possible the following areas: Receptive and expressive communication, information processing capacity. Ability to plan. Use of past experiences to inform future decisions. Mental flexibility—the capacity to switch gears when new information arises. Capacity to use information learned when calm later when under stress. Tendency to be easily over- or underaroused related to sensory input. Sensory responses, i.e., to sounds, touch, visual stimuli, motion, etc. Capacity to understand social cues and social expectations. Wanting connections socially but not understanding how to secure them or maintain them.
Recent stressors	Fill out the RSQ (Recent Stressors Questionnaire) in Box YY. Describe things beyond the usually listed stress events, including the many things that may cause distress to a person but do not show immediate reactions.

Continued

Table 9.1 Components of a developmental–contextual plan: Quick reference—cont'd

Overview	Characterization of IDD, psychiatric diagnoses, and medical diagnoses
Interests and things valued	What usually puts a smile on his or her face? Things that can be done in the future? How can care be built around the things this person values most?
Skill building	These should be aimed at enhancing the ability to use replacements to manage needs and wants, and not simply skills we wish they had. What the person really needs is to stop using negative means in their life. For example, a person may need more help with communication, such as a multimodality system to enhance their current use of words. Or a person may need us to remind them about options each day due to executive function challenges with working memory.
Setting events	Contextual influences are derived from a contextual functional assessment and history taking. Included here are factors related to the person's neurodevelopmental profile, for which it may take a significant amount of time or may not be feasible to teach a replacement skill, such as lifelong sensory sensitivities that we can address but will require time to impact. In addition, specific problems with working memory may necessitate the use of structured schedules and approaches, indicating that the person is unlikely to independently apply coping skills when stressed. Trauma-informed approaches based on "bottom-up" versus "top-down" interventions, such as grounding, relaxation tactics, massage, yoga, swimming, and other similar activities, are also considered. See the example plan for numerous types of setting events often identified for people with ASD and other IDD. Setting events may be viewed as remote antecedents.
Triggers	Immediate antecedents to targeted behaviors, if known, can be addressed with interventions for caregivers or the individual listed here. If it is a "demand," better stated as a "request," should it be described which requests are stressful or upsetting specifically? If sounds or other sensory events are involved, describe these in detail. Any triggers should be defined operationally, so it is clear to everyone when they are occurring.

Table 9.1 Components of a developmental–contextual plan: Quick reference—cont'd

Overview	Characterization of IDD, psychiatric diagnoses, and medical diagnoses
Early warning signs	Critical stage, as incidents rarely occur "out of the blue," and one can observe subtle changes before escalation, such as changes in facial expressions, frowns, decreased eye contact (looking away), pacing, fidgeting, furrowed brows, etc. This is when there is an opportunity to distract or direct the person to something away from the context or offer help before the escalation occurs.
Reactive strategies-hierarchy of responses	The goal is, if all of the above fails, to reduce the intensity and duration of the occurrence and restore calm as rapidly as possible. In most cases, contrary to instinctive interventions, conveying that you are here to help, that you are sorry the person is upset, and that you want to be a support is often more effective than more stern or authoritarian approaches. There are times when very dangerous behavior requires brief physical holds for safety, and even these might be appropriate before the behavior of concern, such as when the person is gesturing in a manner indicating they are about to run into traffic and could die.

Shaping "replacement behaviors" and teaching skills that will reduce aggressive behavior and increase the quality of the person's life are critical in these plans. This requires that we identify teaching methods (i.e., people learn in different ways, as described above). We also need to recognize that a "reward" or motivator for acquiring new skills or behaviors is only a

Example ABC form

ABC Date__________ Time Start_____ Time Over_____ Staff________________

A = All that came before—Tell all that has been going on BEFORE the behavior or symptom

Poor sleep last night?	____YES	____NO
Asked to come out of his room	____YES	____NO
Asked to do ADLs or a chore	____YES	____NO
Pacing or restless?	____YES	____NO
Recently took medication	____YES	____NO
Ate well?	____YES	____NO
Area was loud?	____YES	____NO
Seeking extra food?	____YES	____NO
Asked to wait for bathroom?	____YES	____NO
Asked to wait for anything listed below?	____YES	____NO
Seems tired?	____YES	____NO
Others having a hard time?	____YES	____NO
Using Bathroom a lot?	____YES	____NO

Please elaborate on any and all of the above or other things noted—try to state what was seen versus hypotheses regarding why:

B = Behavior or symptoms or event—Describe what the person did in detail (no labels; describe what you saw and heard, please)—rate the intensity of this: was it minor, moderate, or severe? How long did it last? Was anyone hurt—if so, describe. Note the time the incident started and the time it was clearly over. Who was around at the time, etc.?

C = How was the incident handled? What did the staff do? How did peers react? What tactics exactly were used? Did the person accept staff help? Was an escort required? PHYSICAL HOLD? Restraints? A PRN medication? If so, what? Did the tactics work well or not so well? What activity followed the incident (when it was over—what did he do?)

positive reinforcer if it works to increase the occurrences of these behaviors and that these behaviors are in the service of helping the individual achieve a positive quality of life.

References

Cale, S.I., Carr, E.G., Blakeley-Smith, A., Owen-DeSchryver, J.S., 2009. Context-based assessment and intervention for problem behavior in children with autism spectrum disorder. Behav. Modif. 1552-4167. 33 (6), 707–742. doi:10.1177/0145445509340775.

Carr, E.G., 1999. Positive Behavior Support for People with Developmental Disabilities: A Research Synthesis.

Carr, E.G., Horner, R.H., 1991. Designing positive behavioral support plans: the importance of student, family, and contextual factors. In: Meyer, L.H., Peck, C.A., Brown, L., Eds., Critical Issues in the Lives of People with Severe Disabilities. Paul H. Brookes, Baltimore, MD, pp. 195–210.

Charlot, L., Abend, S., Ravin, P., Mastis, K., Hunt, A., Deutsch, C., 2011. Non-psychiatric health problems among psychiatric inpatients with intellectual disabilities. J. Intellect. Disabil. Res. 55 (2), 199–209.

Gardner, W.I., 2007. Aggression in persons with intellectual disabilities and mental disorders. In: Handbook of Intellectual and Developmental Disabilities. Springer US, Boston, MA, pp. 541–562.

Ghojoghi, F., Abed, E.H., Beigi, M.K., Mahboubi, E., 2024. The effect of token economy on managing aggression in adults and adolescents with mental disorders: a recent review. Middle East J. Rehabil. Health Stud. 2423-4451. 11 (3). Available from: https://brieflands.com/articles/mejrh-142886.pdf. doi:10.5812/mejrh-142886.

Hassiotis, A., Rudra, S., 2022. Behaviours that challenge in adults with intellectual disability: overview of assessment and management. BJPsych Adv. 28 (6), 393–400.

Hassiotis, A., Vegh, B., 2024. New insights on challenging behavior: assessment and treatment across the lifespan. In: Handbook of Psychopathology in Intellectual Disability: Research, Practice, and Policy, pp. 159–185

Hunter, R.H., Wilkniss, S., Gardner, W.I., Silverstein, S.M., 2008. The multimodal functional model–advancing case formulation beyond the "diagnose and treat" paradigm: improving outcomes and reducing aggression and the use of control procedures in psychiatric care. Psychol. Serv. 1939-148X. 5 (1), 11–25. doi:10.1037/1541-1559.5.1.11.

Lehman, B.J., David, D.M., Gruber, J.A., 2017. Rethinking the biopsychosocial model of health: understanding health as a dynamic system. Soc. Pers. Psychol. Compass. 1751-9004. 11 (8). Available from: https://onlinelibrary.wiley.com/journal/10.1111/(ISSN)1751-9004. doi:10.1111/spc3.12328.

Verhoeven, W.M.A., Egger, J.I.M., 2014. Aetiology based diagnosis and treatment selection in intellectually disabled people with challenging behaviours. J. Intellect. Disabil. Diagn. Treat. 2292-2598. 2 (2), 83–93. doi:10.6000/2292-2598.2014.02.02.1.

CHAPTER 10

Concluding remarks: Supporting people with autism and individuals with IDD who display aggressive behavior

The whole is greater than the sum of the parts

Lauren Charlot[a] and Angela Hassiotis[b]
[a]UMass Chan Medical School, Worcester, MA, United States
[b]University College London, London, United Kingdom

Throughout the book, we have provided an integrative overview of how best to assess, understand, and support people with autism and people with intellectual and developmental disability (IDD) who display aggressive behavior. A central theme throughout the book is the value of embedding a *developmental perspective* within a *biopsychosocial* model. This combined framework is essential for understanding why aggressive behavior occurs and for developing intervention plans that move beyond simply reacting to incidents. Instead, clinicians are encouraged to understand the individual's unique neurodevelopmental profile and history and then examine the contextual influences—both historical and current—that shape behavior. This means going beyond typical functional assessments that focus mainly on antecedents and consequences and instead incorporating analyses of *setting events*, including remote, health-related, emotional, and environmental factors that influence vulnerability long before an incident.

Ultimately, reducing or eliminating aggressive behavior requires focusing not only on decreasing aggression but also on *promoting wellness*. This is the purpose of the contextual assessment and treatment (CAT) framework described elsewhere in the book. The CAT approach emphasizes learning as much as possible about how an individual functions when they are at their best, given their neurodevelopmental characteristics. Using well-being as the starting point forms the basis for identifying what disrupts wellness and thereby increases the likelihood of aggression.

Individuals with autism and people with IDD often present with distinctive neurodevelopmental features that can contribute to the risk

Aggressive Behavior of People with Autism and other Intellectual and Developmental Disabilities
https://doi.org/10.1016/B978-0-443-26552-5.00010-2

of aggressive behavior. These features may include difficulties with communication, sensory processing, emotion regulation, executive functioning, and adaptive skills. While these features present challenges, they also come with *strengths* that can be used to support learning and skill development. Understanding both is essential. When assessment reveals that aggressive behavior is associated with a skill deficit—e.g., limited functional communication—the goal becomes to teach or strengthen a replacement skill, utilizing the ways in which the person with IDD learns best, e.g., via visual supports; accommodations for working memory difficulties, slower processing speed, or sensory sensitivities. A personalized approach that respects the individual's developmental profile is key to effective intervention.

It can never be overemphasized that medical issues are a common and often overlooked contributor to aggressive behavior. Many people with autism and other IDD have significant difficulties in accurately reporting internal states, including pain, discomfort, and emotional experiences, although clinical assessments depend heavily on self-report. As a result, undetected or poorly understood medical problems—gastrointestinal issues, dental pain, headaches, and other sources of physical distress—frequently act as *setting events* that increase the likelihood of aggression in response to even small provocations.

The authors caution that clinicians must not assume they have the full picture based solely on interviews. Instead, high-quality assessment should incorporate information from caregivers, historical records, behavioral observations, and situational contexts such as mealtimes or transitions.

A substantial proportion of individuals with autism spectrum disorder (ASD) and people with IDD have epilepsy, including those with poorly controlled seizures. Epilepsy is associated with an increased risk of aggressive behavior through several pathways, ranging from brain abnormalities, seizure type, seizure frequency, and post-ictal states, all of which may influence emotional functioning and behavioral control. Furthermore, antiseizure medications—like all medications—carry side effects, some of which increase discomfort, irritability, or cognitive disruption. To manage these risks, clinicians must carefully consider how best to monitor both wellness and behavioral stability in this context.

The fast-moving field of genetics is greatly associated with autism and other IDDs. The authors describe several examples, focusing on how associated phenotypic features may affect the probability of aggressive behavior. Certain syndromes, such as 22q11.2 deletion, indicate the need for altered medical surveillance because of increased risks (e.g., cardiac issues). Others, like Cornelia de Lange syndrome, have well-described behavioral phenotypes, including high rates of self-injury.

We have highlighted that identifying feasible underlying biological factors helps clinicians anticipate vulnerabilities, tailor environmental supports, and design better preventive interventions. Furthermore, genetic predispositions interact with environmental exposures in dynamic ways, shaping neurodevelopment, stress reactivity, and behavioral expressions across the lifespan.

While much has been written elsewhere about psychiatric assessment in ASD/IDD, we consider a key concept: *aggression has no diagnostic specificity*. Aggressive behavior is not a disorder in itself; rather, it is a "final common pathway" for distress, frustration, or dysregulation—particularly in people with limited means of communicating these internal states. We have started from the principle that clinicians must investigate the underlying pathways leading to aggression instead of assuming aggression reflects a specific psychiatric condition.

Acute psychiatric symptoms can certainly act as *setting events*. However, diagnostic errors are common in this population due to limitations of current diagnostic approaches and the lack of specific tools for individuals with IDD. When clinicians misinterpret anxiety, sensory overload, or distress displayed by a person with autism as symptoms of another disorder—such as bipolar disorder—treatments may exacerbate rather than reduce aggression. Both clinicians and caregivers must be aware of these possibilities.

Polypharmacy is common among individuals with autism and people with IDD, and multiple medications can significantly increase vulnerability to aggressive behavior. Drug interactions, cumulative side effects, and untreated adverse reactions often go undetected—especially for individuals who do not spontaneously report sensations such as akathisia, dizziness, urinary retention, or gastrointestinal discomfort. These physiological stressors may function as potent *setting events* that lower tolerance and increase the probability of aggression, and medication side effects may at times be misinterpreted as attention-seeking behaviors.

Trauma is extremely common in people with autism and individuals with IDD, and we have highlighted how traumatic experiences—physical, sexual, emotional, relational, or environmental—can profoundly affect neurological, psychological, and behavioral functioning. Aggressive behavior is often a manifestation of trauma-related distress, especially in individuals unable to articulate their experiences verbally. Detailed case examples in the chapter illustrate these pathways.

Trauma is also significant for caregivers. Family members and paid staff frequently experience verbal or physical assault and may develop secondary trauma, fear, burnout, or compassion fatigue. These caregiver experiences can influence the emotional climate and become factors that increase individuals' risk of aggressive incidents. Differences in service availability across countries, states, and healthcare systems further complicate access to trauma-informed support for both individuals and caregivers.

Charlot expands on a developmentally informed PBS model, integrating the CAT framework. This approach prioritizes understanding contextual influences on the likelihood of aggression. The emphasis is on prevention: identifying when and why a person is more vulnerable and designing supports that reduce triggers and strengthen protective factors.

Contextual assessments involve systematic data collection, often conducted by multidisciplinary teams investigating medical, psychiatric, neurological, environmental, and interpersonal contributors to behavior. A 24-h interval data sheet, used extensively in the UMass specialty psychiatric unit, helps identify patterns related to illness, discomfort, sleep, environmental demands, social interactions, and positive well-being indicators. Such data inform concise, user-friendly *Developmental-Contextual Assessment and Treatment Plans* that highlight actionable strategies, early warning signs, and pathways for further assessment.

Case examples illustrate the application of this approach, and a range of contextual tools can help in the identification not only of risks but also of the conditions under which individuals are happiest and most well.

As clinicians and researchers, we acknowledge the complexity of supporting people with autism and people with IDD who present with aggressive behavior. Effective input requires developing a deep understanding of each person's unique neurodevelopmental and biopsychosocial profile. This is inherently a multidisciplinary, collaborative process—one that truly "takes a village." By integrating developmental science, contextual assessment, trauma-informed care, careful medical and psychiatric evaluation, and strengths-based PBS approaches, clinicians can profoundly improve the lives of individuals who are among the most underserved in society.

Index

Note: Page numbers followed by "*f*" indicate figures, "*t*" indicate tables, "*b*" indicate boxes.

Printed and bound by CPI Group (UK) Ltd, Croydon, CR0 4YY
14/09/2026
02216297-0001